AI in Manufacturing and Green Technology

Green Engineering and Technology: Concepts and Applications

Series Editors:
Brujo Kishore Mishra
GIET University, India and Raghvendra Kumar, LNCT College, India

Environmental conservation is an important issue these days for the whole world. Different strategies and technologies are used to save the environment. Technology is the application of knowledge to practical requirements. Green technologies encompass various aspects of technology which help us reduce the human impact on the environment and create ways of sustainable development. Social equability linked to this book series will enlighten the green technology in different ways, aspects, and methods. This technology helps people to understand the use of different resources to fulfill their needs and demands. Some points will be discussed as the combination of involuntary approaches, government incentives, and a comprehensive regulatory framework to encourage the diffusion of green technology. However, least developed countries and developing states of small islands require unique support and measures to promote green technologies.

Green Innovation, Sustainable Development, and Circular Economy
Edited by Nitin Kumar Singh, Siddhartha Pandey,
Himanshu Sharma, and Sunkulp Goel

Green Automation for Sustainable Environment
Edited by Sherin Zafar, Mohd Abdul Ahad,
M. Afshar Alam and Kashish Ara Shakeel

AI in Manufacturing and Green Technology: Methods and Applications
Edited by Sambit Kumar Mishra, Zdzislaw Polkowski,
Samarjeet Borah, and Ritesh Dash

Green Information and Communication Systems for a Sustainable Future
Edited by Rajshree Srivastava, Sandeep Kautish, and Rajeev Tiwari

For more information about this series, please visit: https://www.crcpress.com/Green-Engineering-and-Technology-Concepts-and-Applications/book-series/CRCGETCA

AI in Manufacturing and Green Technology
Methods and Applications

Edited by
Sambit Kumar Mishra, Zdzislaw Polkowski,
Samarjeet Borah, and Ritesh Dash

CRC Press
Taylor & Francis Group
Boca Raton London New York

CRC Press is an imprint of the
Taylor & Francis Group, an **informa** business

First edition published 2021
by CRC Press
6000 Broken Sound Parkway NW, Suite 300, Boca Raton, FL 33487-2742

and by CRC Press
2 Park Square, Milton Park, Abingdon, Oxon, OX14 4RN

© 2021 Taylor & Francis Group, LLC

CRC Press is an imprint of Taylor & Francis Group, LLC

Library of Congress Cataloging-in-Publication Data
Names: Mishra, Sambit Kumar, editor. | Polkowski, Zdzislaw, editor. |
Borah, Samarjeet, editor. | Dash, Ritesh, editor.
Title: AI in manufacturing and green technology : methods and applications /
edited by Sambit Kumar Mishra, Zdzislaw Polkowski, Samarjeet Borah,
and Ritesh Dash.
Description: First edition. | Boca Raton, FL : CRC Press, 2020. |
Series: Green engineering and technology | Includes bibliographical references
and index.
Identifiers: LCCN 2020014299 (print) | LCCN 2020014300 (ebook) |
ISBN 9780367895655 (hardback) | ISBN 9781003032465 (ebook)
Subjects: LCSH: Manufacturing processes—Data processing. |
Green technology—Data processing. | Artificial intelligence.
Classification: LCC TS183 .A44 2020 (print) | LCC TS183 (ebook) |
DDC 670.285/63—dc23
LC record available at https://lccn.loc.gov/2020014299
LC ebook record available at https://lccn.loc.gov/2020014300

ISBN: 978-0-367-89565-5 (hbk)
ISBN: 978-0-367-53740-1 (pbk)
ISBN: 978-1-003-03246-5 (ebk)

Typeset in Times
by codeMantra

Contents

Preface

This book prioritizes working towards sustainability of the environment, employing engineering aspects along with green computation and applying various aspects of modern education as well as solutions. Practically, it visualizes specific aspects of artificial intelligence in manufacturing and green technology, covering the implementation of renewable resources enhancing business activities. As the term green technology is focused on preserving the resources and environment along with controlling the negative impact from human activities with sustainable development, obviously the solutions take into account social, economic as well as environmental concerns. The concept of artificial intelligence is used to optimize manufacturing supply chains and enhance energy efficiency, thereby helping companies to anticipate market changes. In this book, implementation has also been emphasized via artificial intelligence in resources with changing technologies where the initiation to process automation has been preserved. In addition to different parameters to measure the efficiency, various characteristics linked to high-end manufacturing activities have been focused. Considering the present situation, the concept of green data center based on storage, management and dissemination of data with control strategies has been placed in this book. It is understood that the concept of green computing, which links to eco-friendly implementation of systems along with resources, is associated with the computing devices in such a way to minimize any adverse impact on the environmental issues in the IT industry. In such situation, along with green technology, the entire environment is looking forward to more energy-efficient mechanisms, managed security services, cloud security solutions at one place by offering equivalent along with cloud management platforms. Similarly focusing on renewable energy—very specific energy which is collected from renewable resources—are commonly wind, sunlight, as well as geo-thermal heat, which often has a provision in different major areas like generation of electricity, transportation and rural energy services. Also, with the significant role of green computation in the sensor-enabled IoT, smart application enables the sensors in the situation where faster energy depletion is responsible for efficient functioning of the sensors. As such, a specific technique is needed to protect the sensors in IoT. In such cases, meta-heuristic techniques can be the better solution to solve similar types of situations with the near-optimal solutions.

MATLAB® is a registered trademark of The MathWorks, Inc. For product information, please contact:
   The MathWorks, Inc.
   3 Apple Hill Drive
   Natick, MA 01760-2098, USA
   Tel: 508-647-7000
   Fax: 508-647-7001
   E-mail: info@mathworks.com
   Web: www.mathworks.com

Editors

Sambit Kumar Mishra has more than 22 years of working experience in different AICTE-approved institutions. He has made more than 29 publications in different peer-reviewed international journals. He is editorial board member of different peer-reviewed indexed journals. Presently, he is associated with Gandhi Institute for Education and Technology, Baniatangi, Bhubaneswar, Odisha, India.

Zdzislaw Polkowski is presently associated with Wroclaw University of Economics and Business, Poland. He holds a Ph.D. in Computer Science and Management from Wroclaw University of Technology, postgraduate degree in Microcomputer Systems in Management from University of Economics in Wroclaw, and postgraduate degree in IT in Education from Economics University in Katowice. He obtained his Engineering degree in Industrial Computer Systems from Technical University of Zielona Gora. He has published more than 55 papers in journals, 15 conference proceedings, and more than 8 papers in journals indexed in the Web of Science. He has served as a member of Technical Program Committee in many international conferences in Poland, India, China, Iran, Romania, and Bulgaria.

Samarjeet Borah is currently working as Professor in the Department of Computer Applications, Sikkim Manipal University (SMU), Sikkim, India. Dr. Borah handles various academics, research, and administrative activities. He is also involved in curriculum development activities, board of studies, doctoral research committee, IT infrastructure management, etc. along with various administrative activities under SMU. Dr. Borah is involved with various funded projects in the capacity of Principal Investigator/Co-principal Investigator. The projects are sponsored by agencies like AICTE (Government of India), DST-CSRI (Government of India), Dr. TMA Pai Endowment Fund, etc. He is associated with ACM (CSTA),

IAENG, and IACSIT. Dr. Borah has organized various national and international conferences in SMU. Some of these events include ISRO Sponsored Training Programme on Remote Sensing and GIS, NCWBCB 2014, NER-WNLP 2014, IC3-2016, IC3-2018, and IC3-2020. He is also associated with various other conferences in the capacity of steering committee member, TPC member, editorial board member, volume editor, and reviewer. Dr. Borah is involved with various book volumes and journals of repute in the capacity of editor/guest editor/reviewer such as *IJHISI, IJGHPC, IJVCSN, IJIPT, IJDS, IJBM, WRITR*, and *IEEE Access*. He is editor-in-chief of the series *Research Notes on Computing and Communication Sciences*, Apple Academic Press (exclusive worldwide distribution by CRC press).

Ritesh Dash is presently associated with Christian College of Engineering, Bhilai, Chhatisgarh. He has made a good number of publications in various indexed and Scopus journals.

Contributors

T. Badapanda
Department of Physics
CV Raman College of Engineering
Bhubaneswar, India

Urmila Bhanja
Department of Electronics and
    Communication Engineering
Indira Gandhi Institute of
    Technology
Sarang, India

S.S. Bishoyi
Department of Mathematics
Christian College of Engineering and
    Technology
Bhilai, India

Samarjeet Borah
Department of Computer Science and
    Applications
Sikkim Manipal Institute of
    Technology
Gangtok, India

Ritesh Dash
Department of Electrical
    Engineering
Christain College of Engineering
    and Technology
Bhilai, India

Nirupama Deep
Department of Electronics and
    Communication Engineering
Gandhi Institute for Education and
    Technology
Baniatangi, India
and
Department of Electronics and
    Communication Engineering
Centre for Advanced Post Graduate
    Studies, BPUT
Rourkela, India

Jayanta Kumar Kar
Department of Electrical Engineering
National Institute of Technology
Silchar, India

G. Ajith Kumar
Department of Mechanical Engineering
School of Engineering
Cochin University of Science and
    Technology
Cochin, India

Pati Sushil Kumar
Department of EE
Gandhi Institute for Education and
    Technology
Baniatangi, India

Antaryami Mishra
Department of Mechanical Engineering
Indira Gandhi Institute of Technology
Sarang, India

S.R. Mishra
Department of Chemistry
Gandhi Institute for Education and
    Technology
Bhubaneswar, India

Sambit Kumar Mishra
Department of Computer Science and
    Engineering
Gandhi Institute for Education and
    Technology
Bhubaneswar, India

Narayan Chandra Nayak
Department of Mechanical Engineering
Indira Gandhi Institute of Technology
Sarang, India

Chinmayee Panda
Department of Electronics and
    Communication Engineering
Indira Gandhi institute of Technology
Sarang, India

Sudhanshu Bhushan Panda
Department of Mechanical Engineering
Indira Gandhi Institute of Technology
Sarang, India

Binaya Kumar Panigrahi
Department of Civil Engineering
Gandhi Institute for Education and
    Technology
Baniatangi, India

S. Parida
Department of Physics
CV Raman College of Engineering
Bhubaneswar, India

Suraj Pattanaik
Department of ECE
Gandhi Institute for Education and
    Technology
Baniatangi, India
and
Department of ECE
Centre for Advanced Post Graduate
    Studies, BPUT
Rourkela, India

Zdzislaw Polkowski
Department of Information and
    Communication Technology
Wroclaw University of Economics and
    Business
Wroclaw, Poland

Devidarshinee Pradhan
Department of Electrical Engineering
College of Engineering and Technology
Bhubaneswar, India

Sovit Kumar Pradhan
Department of Electrical Engineering
National Institute of Technology
Silchar, India

Suman Sourav Prasad
Department of Master in Computer
    Applications
Ajay Binay Institute of Technology
Cuttack, India

Pradyumna Kumar Sahoo
Electrical Engineering
Gandhi Institute for Education and
    Technology
Baniatangi, India

Prateek Kumar Sahoo
Department of Electrical Engineering
SOA University
Bhubaneswar, India

Sudhansu S. Sahoo
Department of Mechanical
  Engineering
College of Engineering and
  Technology
Ghatikia, India

Soumya Ranjan Satapathy
Department of Civil Engineering
Gandhi Institute for Education and
  Technology
Baniatangi, India

Prasanta Kumar Satpathy
Department of Electrical Engineering
College of Engineering and Technology
Bhubaneswar, India

Satpathy Suchismita
Department of Electronics and
  Telecommunications
CV Raman Polytechnics
Bhubaneswar, India

Sarat Chandra Swain
School of Electrical Engineering
KIIT University
Bhubaneswar, India

Sanju Thomas
Department of Mechanical Engineering
School of Engineering, Cochin
  University of Science and
  Technology
Cochin, India

P.K. Tripathy
Department of Mathematics and
  Science
Utkala Gourav Madhusudan Institute of
  Technology
Rayagada, India

1 Concept of Green Computation Linked to Cloudlets
A Case Study

Zdzislaw Polkowski
Wroclaw University of Economics and Business

Suman Sourav Prasad
Ajay Binay Institute of Technology (Affiliated
to Biju Patnaik University of Technology)

Sambit Kumar Mishra
Gandhi Institute for Education and Technology
(Affiliated to Biju Patnaik University of Technology)

CONTENTS

1.1 INTRODUCTION

Being associated with the internet-of-things environment, it is obvious that the loosely connected devices are connected through heterogeneous networks. The main intention in this case is to collect and process data from the devices linked with internet of things in order to mine and detect patterns, perform predictive analyses or optimization, and ultimately provide better decisions within the stipulated

1

time. Data in such cases can be accumulated either in big stream or as big data. In some cases, the transient data can be captured constantly from smart devices linked with internet of things, and in most cases, the persistent data and knowledge can be stored and archived in centralized cloud storage environments. As soon as data are collected and aggregated from a virtual network along with smart devices, they are linked with the cloud servers. In such case, the cloud computing offers a solution at the infrastructure level that supports processing of big data. It enables highly scalable computing platforms that can be configured on demand to meet constant changes of application requirements in a pay-per-use mode, thus reducing the investment necessary to build the desired analytics application. As mentioned previously, this perfectly matches the requirements of big data processing when data are stored in a centralized cloud storage. By the way, as compared with the cloud, the fog provides services with a faster response. Accordingly, the fog computing is treated as a better option to enable the facilities of internet of things to provide efficient and secure services. The combination of internet of things with cloud in the long run is a better option to overcome all the issues related to storage as well as data processing in cloud. Basically, it simplifies the mechanism of gathering the data in virtual servers and provides low-cost installation and integration for complex data processing as well as deployment. Also, these are subsequently analyzed to determine decisions regarding implementation mechanisms. Linking these data with cloud usually requires excessively high network bandwidth. To overcome these issues, fog computing can be adopted whose application is very similar to cloud. It provides services to users and is based on providing data-processing capabilities and storage locally in fog devices instead of sending them to the cloud. Each fog node hosts local computation, networking, and storage capabilities.

1.2   REVIEW OF LITERATURE

Vats et al. [1] in their work have focused on the effectiveness of cloud computing in analyzing the performance of computational platforms and data centers. They observed that the centralized models will perform better with the adequate number of connected devices.

Hany et al. [2] in their work have focused on the basic features of fog computing linked to mobility, wireless accessing, streaming, and real-time applications. They also observed the effectiveness of these applications in different internet-of-things applications and services. As such, sometimes implementing similar techniques becomes challenging in terms of scalability, complexity, dynamicity, and heterogeneity.

Yeboah-Boateng et al. [3] in their study have discussed about the mechanisms of cloud and its link with fog computing to handle existing situations. They tried to mobilize the services by providing accessibility of data in remote locations.

Satyanarayanan [4] primarily focused on the effectiveness of fog computing linked to different distributed information collection points. He observed that the synthesis of fog and cloud computing has a great impact on private and public organizations.

Rajkumar et al. [5] in their study described the application of internet of things in various sectors such as process and discrete manufacturing industries, energy and power industries, connected cars, services, connected towns, and transportation.

Adams [6] discussed the utilities and opportunities associated with the integrated mechanism of fog and cloud computing. According to Adams, the efficiency can be increased by enhancing material utilization with scalability and flexibility.

In their work, Botta et al. [7] tried to focus on the support of fog and cloud computing based on real-time associations between internet-of-things tools to minimize latency constraints in data processing and analysis.

Ahmed and Rehmani [8] discussed the impacts of fog and cloud computing with their implementation in different applications related to internet of things. They observed that integration of fog and cloud computing will be definitely beneficial toward agility and data safety.

Aazam and Huh [9] proposed the architecture layer linked to fog computing to minimize resources used in servers. They observed that, in general, the probability-based model can be used to analyze the characteristics of fog computing along with the requirement of resources.

Wang et al. [10], in their study, provided the architectural concept of fog computing and discussed the applications and mechanisms of fog computing nodes. They also differentiated the performance of conventional cloud computing schemes with the application of internet of things in smart grids.

In Ref. [11], it has been observed that fog computing sometimes faces challenges while linked with edge computing, and also the issues can be highlighted while the applications are associated with centralized distributed platforms.

Ab Rashid and Ravindran [12] discussed the capabilities of cloud and the ability of processing computational power in the edge device over the network. They observed the vital issues and challenges of cloud computing in such scenario while associated with resource allocation and provisioning, network congestions, and privacy and security management.

Redowan and Rajkumar [13] described the localized and decentralized nature of fog. Practically, it offers the cloud users a choice on how many resources to use and when to release them, which leads to better resource allocation and distribution. So, the cost becomes optimal and the system proves to be cost-efficient.

Fatemeh et al. [14] studied the centralized distribution of computing resources. They discussed the bottlenecks in applications of fog computing, in which only the trimmed data are sent to the network through edge devices making it efficient and elegant with uncongested interconnectivity. They also focused on different energy optimization schemes which, according to their results, reduce the energy consumption. As such, the fog computing can be much beneficent to the centralized domain.

Mohammad and Vatanparvar [15] in their work discussed the resources and services linked to fog computing. According to their experimentation, the decentralized fog architecture provides data proximity to end users than cloud-centralized architecture, and it also offers better location resource awareness, powerful processing computational capabilities, and real-time decision-making in a ubiquitous computing environment.

Hassan et al. [16] studied how virtual platforms handle devices like mobiles to store data and speed up their computing power, and observed the limitations of the cloud while adopting fog computing. They also observed how embedded intelligence helps the devices to make decisions at the edge of the network and improve data throughput in real time.

Ivan and Wen [17] discussed the Cisco fog model linked to all connected devices. They described the importance of cloud and edge paradigm in technology shift, latency enhancement, location awareness, and improving quality of services.

Bhagyashree and Geeta [18] explained the goals of fog computing and focused on adoption of fog computing to enhance potency and scale back the amount of data transported to cloud.

In their study, Ab Rashid and Ravindran [19] focused on cloud domain along with computing paradigm. It efficiently reduces the latency issues. According to their observation, latency is an important aspect that enriches the communication process, considering the real-time sensitive decision-making.

1.3   CHALLENGES OF CLOUD OF THINGS

The combination of virtualized environment and internet of things practically has enormous advantages. It helps to organize the resources linked to internet of things along with a provision of cost-effective and sustainable internet-of-things services. Also, it manages with a provision of deployment of processed and complex data with proper integration. Sometimes it invites new challenges to the system linked to internet of things, as some of these cannot be addressed by the traditional centralized cloud computing architecture. As observed, the centralized cloud approach is not appropriate for applications linked to internet of things where operations are time-sensitive as well as lack internet connection.

1.4   ADOPTION OF FOG COMPUTING FOR ELASTIC COMPUTATION AND STORAGE

As the name fog computing implies with its capabilities linked to computing and networking, solutions may be obtained considering the computing architecture distributed geographically along with the provision of elasticity. It also possesses system-level horizontal architecture for distribution of resources and services of computing, storage control, and networking. Basically, it expands its architectural mechanisms to the edge of the network. Although the fog and the cloud acquire similar resources and implement very similar mechanisms with signified attributes, fog computing is more beneficial for devices linked with internet of things. Of course, it has the capability to support real-time services with a provision of distributed computing and storage resources to large and widely distributed applications. It allows the collaboration of different physical environments and infrastructures among multiple services.

1.5 PREVAILING SECURED MECHANISM WITH RELIABILITY

Issues associated with the protocols can be proved secured by implementing fog computing over a distributed system. In practical applications, fog computing sometimes faces challenges with the same. The major security issue in such case is the authentication which is related to public key infrastructure. In such cases, potential solutions to this authentication problem can be achieved through measurement-based methods to find the rogue devices and reduce authentication cost. As the fog computing links with a large number of geographically distributed devices and connections, along with the current mechanisms, packet reliability and event reliability are also major concerns. So, it is highly essential to minimize the information accuracy problem along with redundancy.

1.6 ENERGY MINIMIZATION

As the environment supporting the fog computing is associated with various fog nodes with deployment capabilities, it is required to consider the issues of energy efficiency as well as centralized computation even in virtual environment. So, actually it is a big challenge. Again, in this case, it is required to work on the mechanism behind power consumption along with consideration of delay while implementing fog computing system. Also, the distributed solution toward this approach is essential to focus on performance and scalability issues (Figures 1.1 and 1.2).

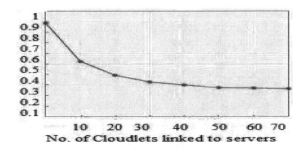

FIGURE 1.1 Number of cloudlets versus cost of cloudlets.

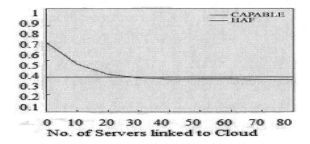

FIGURE 1.2 Servers linked to cloud versus cost of cloudlets.

TABLE 1.1

Cost of Cloudlets Linked to Servers in Cloud

Serial Number	Number of Cloudlets per Server	Number of Servers Linked to Cloud	Processing Cost of Cloudlets Linked to Servers (ms)
1.	20	20	0.51
2.	30	30	0.43
3.	40	40	0.44
4.	50	50	0.39

1.7   APPLICATION OF PARTICLE SWARM OPTIMIZATION TOWARD CATEGORIZING AND PAIRING CLOUDLETS

Basically, particle swarm optimization (PSO) is a bionic intelligent optimization algorithm and a stochastic optimization approach. It signifies the information-sharing mechanism to achieve the optimal solution. Each particle associated with the population represents a possible solution of the problem to be optimized (Table 1.1).

Algorithm 1: Application of PSO toward Pairing Cloudlets

Step 1: Initialize parameters, that is, size of data items and cloudlets, and set the iteration's maximum number 80, the population size 100, inertia weight range, and learning factors.

Step 2: Initialize population randomly, and set the initial positions and velocities of the population randomly.

Step 3: Calculate the fitness value of all the cloudlets.

Step 4: For each cloudlet c_i, compare the fitness function value and best position. If the value of c_i is smaller, replace it as best position linked to all the cloudlets.

Step 5: For each cloudlet c_i, compare the fitness function value with the global best position. If the value is smaller, then it will be recorded as the current global best position.

Step 6: For each cloudlet, update its velocity and position.

Step 7: Decide whether the iteration achieves the maximum value. Else, again follow Step 3 for the next iteration and achieve the optimal result.

Algorithm 2: Identification of Location of Devices and Clusters

Step 1: Identify the location of devices and clustered link to central location, dl_j.

Step 2: Obtain the data items and cloudlets linked to central location, cdl.

Step 3: Initiate the process and formulate the mechanisms associated with the cloudlets for $i = 1$ to cdl do

dl_i = null
for $j = 1$ to n do
achieve the location of dl_i.
Step 4: Obtain the distance between cdl and dl_i.
Step 5: If (distance < threshold, th), then
Step 6: Link dl_i with cdl.
Step 7: If (dl_i < threshold, th), update the cloudlets linked to cdl.

Algorithm 3: Assessment of Cloudlets by Pairing

Step 1: Obtain the cloudlets linked to central location, cdl.
Step 2: Make the pairing of cloudlets with databases linked to central location, cld_i.
Step 3: Determine the effectiveness of cloudlets in each pair and obtain the threshold value, th. Step 4: Obtain the size of the cloudlets and pair linked with the central location.
for $i = 1$ to cdl do
link the cloudlet, cld_i to cdl.
Step 5: Apply sorting mechanisms to attain cloudlets and arrange in increasing order.
Step 6: Link the threshold value th to cld_i.
Step 7: While $th > 0$ do
for $j = 1$ to n do
if cld_i has no pair of cloudlets, then link cld_i with the central location again.

1.8   DISCUSSION AND FUTURE DIRECTION

It is observed that fog computing somehow fills the gap between the cloud and end devices through initiating computing, storage, networking, and data management. However, the decision-making and data management not only occur in the cloud. The concept of cloud computing practically associates with large data centers; in contrast, the fog computing implements through small servers, routers, switches, gateways, etc. As minimum storage space is required for fog computing, in this regard, the necessary hardware can be located closer to the users.

While observing the dynamism among the cloudlets linked to green energy, it is clear that some of the cloudlets will be approachable towards green energy but, on the other hand, may seek grid energy. Somehow the green energy is practically associated with cloudlet. In this regard, it must be completely utilized to increase the efficiency and minimize the discrepancies.

1.9   CONCLUSION

Practically, it is essential to concentrate on facilitating allocation and linking of big data. In such cases, to minimize the maintenance and operational costs of distributed cloudlets, it is required to energize the cloudlets applying green technologies.

Accordingly, application of green technologies minimizes the operational cost of cloudlets. Considering the basic concept of green energy, the cloudlets should be provided with green technologies or green computation to satisfy the current needs.

REFERENCES

1. Vats, K., Sharma, S., and Rathee, A. (2012). A review of cloud computing and e-governance. *International Journal of Advanced Research in Computer Science and Software Engineering*, 2: 185–213.
2. Atlam, H.F., Walters, R.J., and Wills, G.B. (2018). Fog computing and the internet of things: A review. *Big Data and Cognitive Computing*, 2: 1–18.
3. Yeboah-Boateng, E.O., and Essandoh, K.A. (2014). Factors influencing the adoption of cloud computing by small and medium enterprises in developing economies. *International Journal of Emerging Science and Engineering*, 2(4): 13–20.
4. Satyanarayanan, M. (2017). The emergence of edge computing. *Computer*, 50(1): 30–39.
5. Buyya, R., Yeo, C.S., and Venugopal, S. (2008). Marketoriented cloud computing: Vision, hype, and reality for delivering it services as computing utilities. In *10th IEEE International Conference on High Performance Computing and Communications, HPCC'08*. IEEE, 2008, Bangalore, India, pp. 5–13.
6. Adams, F. (2017). OpenFog reference architecture for fog computing. https://knect365. com/Cloud-enterprise-tech/article/0fa40de2-6596-4060-901d-8bdddf167cfe/openFog-reference- architecture-for-Fog computing.
7. Botta, A., et al. (2016). Integration of cloud computing and internet of things: A survey. *Future Generation Computer Systems*, 56: 684–700.
8. Ahmed, E., and Rehmani, M.H. (2017). Mobile edge computing: Opportunities, solutions, and challenges. *Future Generation Computer Systems*, 70: 59–63.
9. Aazam, M., and Huh, E.-N. (2015). Fog computing micro datacenter based dynamic resource estimation and pricing model for IoT. In *IEEE 29th International Conference on Advanced Information Networking and Applications*, Gwangju, South Korea, pp. 687–694.
10. Wang, P., Liu, S., Ye, F., and Chen, X. (2018). A fogbased architecture and programming model for IoT applications in the smart grid, arXiv.org.
11. OpenFog Architecture Overview. OpenFog Consortium Architecture Working Group. Accessed on Dec. 7, 2016.
12. Ab Rashid, D., and Ravindran, D. (2016). Survey on scalability in cloud environment. *International Journal of Advanced Research in Computer Engineering & Technology*, 5(7): 440–445.
13. Redowan, M., and Rajkumar, B. (2016). Fog computing: A taxonomy, survey and future directions. In Di Martino, B., Li, K.-C., Yang, L.T., Esposito, A. (Eds.) *Internet of Everything: Algorithms, Methodologies, Technologies and Perspectives*. Springer, Singapore. doi:10.1007/978–981-10-5861-5_5.
14. Fatemeh, J., Kerry, H., Robert, A., Tansu, A., and Rodney, S.T. (2016). Fog computing may help to save energy in cloud computing. *IEEE Journal on Selected Areas in Communications*, 34(5): 1–12. doi:10.1109/JSAC.2016.2545559.
15. Mohammad, A.A., and Vatanparvar, K. (2015). Energy management-as-a-service over fog computing platform. *IEEE Internet of Things Journal*, 3: 161–169. doi:10.1109/JIOT.2015.2471260.
16. Hassan, M., Xiao, M., Wei, Q., and Chen, S. (2015). Help your mobile applications with fog computing. doi:10.1109/SECONW.2015.7328146.

17. Ivan, S., and Wen, Sh. (2014). The fog computing paradigm: Scenarios and security issues. In *Federated Conference on Computer Science and Information Systems, FedCSIS 2014*, Warsaw, Poland, pp. 1–8. doi:10.15439/2014F503.

18. Bhagyashree, D., and Geeta, T. (2017). Review on fog: A future support for cloud computing. *International Journal of Research in Science & Engineering*, 3(3): 345–350.

19. Ab Rashid, D., and Ravindran, D. (2019). Fog computing: An efficient platform for the cloud-resource management. *International Journal of Emerging Technologies and Innovative Research*, 6(2): 7–12.

2 Scheduling Tasks in Virtual Machines Using Ant Colony Optimization Technique

Samarjeet Borah
Sikkim Manipal Institute of Technology

Sambit Kumar Mishra
Gandhi Institute for Education and Technology
(Affiliated to Biju Patnaik University of Technology)

CONTENTS

2.1 INTRODUCTION

In the distributed computing environment, the cost of processing depends upon the size of cluster and grid. In distributed computing, the architecture is more reliable and elastic. In cloud computing, which can be considered as a more prominent distributed paradigm, the hardware and software are facilitated through internet. As such, cloud computing may be linked with the features of grid computing and virtualization. The cloud gathers the bulk of computer resources quite similarly to distributed system with the purpose of addressing the user's request. It may also provide the path for capturing the computing power and storage capacity so as to provide the required service at low prices. The virtualization technology may also be associated with a large number of servers, data stores, networks, and software that virtualize a single system into a number of virtual systems. The main aim of scheduling the virtual machines is to allocate the processing cores of a host to virtual machine toward optimization. Scheduling in cloud computing in general case may be linked to NP-hard problem, and therefore, it may lead to complexity while obtaining optimal solutions.

2.2 REVIEW OF LITERATURE

Ristenpart [1] discussed authentication and authorization linked to virtual host auditing. He highlighted on the problems related to users alongside the hypervisor and mostly to virtualization security. Sabahi [2] provided the implementation of virtual machines using hypervisor. He also discussed how an IDS within the hypervisor may efficiently detect a similar IDS while running on a guest operating system. Because the guest operating system cannot monitor events in the cloud, it is now possible for the guest operating systems to watch virtual machine events. Gallagher [3] presented residual physical representation of knowledge, that is, how critical data are reconstructed and guarded after several changes and modifications. Fletcher [4] described how IaaS providers work on storage capacity coupled with registration process to implement cloud services. Kalra and Singh [5] proposed the metaheuristic techniques to schedule the tasks within the area of cloud computing and grid computing. They observed that the techniques are proved to realize near-optimal solution and should be compared with NP-hard problem. Tawfeek et al. [6] focused on issues related to task scheduling in cloud computing environment. They observed it as NP-hard optimization problem to get the near-optimal solution of this massive size problem. Madhvi and Sowmya Kamath [7] proposed a hybrid approach of task scheduling. They implemented the hybrid task scheduling algorithm by increasing the number of cloudsets and virtual machines. Kruekaew and Kimpan [8] discussed the essential concepts of metaheuristic approach for scheduling the virtual machines in cloud environment. During their study, they analyzed the algorithm and implemented the scheduling concepts within the cloud environment. Nishant et al. [9] implemented an algorithm related to the workloads among nodes of a cloud using ant colony optimization technique. They applied ant colony optimization technique over cloud network systems so as to take care of the load over the nodes and to detect the overloaded nodes with their threshold values. Calheiros et al. [10] discussed allocation of virtual machines along with scheduling techniques. They followed the strategies of allocation of free resources considering the space and time sharing scheduling strategies. They also observed the concurrent execution of virtual machines. Chen et al. [11] described hybrid genetic algorithm to obtain fitness parameters based on memory size of corresponding virtual machines. It has been provided with high resource utilization and accordingly decreased the number of physical machines for efficient energy conservation. Liu et al. [12] utilized ant colony optimized virtual machine placement approach to decrease the number of physical machines. It has been observed that the approach is able to achieve more efficient resource usage with a more number of virtual machines. Christina et al. [13] discussed the techniques to link virtual machines to relevant physical machines. In this regard, they also proposed an approach to minimize the energy consumption and migration rate. Lin [14] proposed techniques associated with resource scheduling based on time, cost, and energy. In this work, it is shown that how to assist cloud users in selecting suitable scheduling algorithms based on their service needs. In their study, Liu et al. [15] used directed acyclic graph-based scheduling algorithm to schedule multiple tasks in heterogeneous cloud computing environments. It has been observed that preprocessing of cloud resources may be completed before task scheduling. In his work, Koomey [16] focused on consumption of computing energy and performance of the system. It has been observed that the cooling cost and the consumption of computing energy can be optimized by considering alternative scheduling mechanisms linked to data centers. Banerjee et al. [17] described a specific mechanism to rank the servers based on their maximum utilization along with consumption of power and recirculation of temperature. In this regard, they focused on thermal sensors for real-time monitoring. Wang et al. [18] focused on scheduling mechanisms to simulate the thermal maps of sorted servers to monitor the temperatures. It is seen that as compared with other thermal maps, the thermal map implemented in their work was primarily linked to actual node temperatures. Maruyama et al. [19] considered a mechanism to minimize the consumption of electricity required for cooling the required equipment. They implemented different heat control methods in order to eradicate hot spots.

2.3 CONCEPTUALIZATION OF GREEN SCHEDULING AND COMPUTATION

It is obvious that computation in virtual environments is more intensive as it takes much less time for execution. In fact, a sequence of low-capacity systems may be more profitable instead of acquiring high-capacity systems. Also, the highly signified systems sometimes may not utilize all resources at a time. Thus, improvement in performance may be observed by implementing the scheduling concepts of green computation in virtual platforms. Basically, the optimization process is carried out by operating system using the scheduling techniques based on green computation. The green compiler which is nothing but a system software and translator reshapes the transformations and also tries to maintain the cost of database for each instruction. The database then parses the codes and initiates regeneration. This technique may also ensure optimality while processing queries. Sometimes, while linking and initiating tasks in cloud platforms, the compiler implements the method of acquiring the cloud services at host level. But the consistency may not be maintained due to lack of facilities of virtual systems with a strong internet connection. The green computation technique utilizes computing resources very effectively. It also helps in making data centers ecofriendly as it works for optimal utilization of information technology. Hence application of green computing in data centers may be helpful to enhance productivity and to provide support for generating a pollution-free environment. In this regard, strategies based on green computing may be more beneficial for data centers and growth of IT industries.

2.4 PRACTICAL IMPLICATIONS TOWARD DATA ACCUMULATION

Much experimentation has been conducted to minimize the acquaintance of energy related to data storage in virtual environments. Practically, the experiments were intended to evaluate the energy associated with virtual machines. The data centers linked to clouds should definitely be process oriented and schedulable toward systems. While experimenting the design applications of different data centers, it may be observed that the permission to access the tasks along with the scheduled data sometimes be resynchronized dynamically and centered to nominate the key features of virtual machines. As well, data may be collected and validated by applying various life-cycle strategies. Applications of green computing sometimes can be measured for their internal consistency ranging from 0 to 9, with the value of absorption be 0.986. The facilitation of greening the data centers practically is verified based on associated metrics or retrieved data.

2.5 PRACTICAL AND RESEARCH IMPLICATION

Broadly speaking, the approach of green technology relates to the life-cycle strategies associated with green design, production, operation, and disposal with an effective approach toward obtaining a solution. For example, the system based on green computing yields minimum emission of carbon dioxide and, at the same time, has enhanced efficiency. The implication is also associated with the strategies of green computing and can be utilized as a benchmark by IT-based industries. IT-based industries can apply green practice initiatives in their data centers to make them more eco-friendly.

2.6 IMPLEMENTATION OF ANT COLONY OPTIMIZATION TECHNIQUES

The ant colony optimization technique is a metaheuristic approach and is based on population and probability. In particular, it is motivated from the behavior of ants. It is implemented to obtain the optimized solution for the computational problems and may be quite competent to find the shortest path through pheromones. The intensity of pheromones can be enhanced by increasing the number of ants passing through the path. In such cases, the shortest path may be obtained through the pheromones trail.

2.7 CRITERIA FOR IMPLEMENTATION

Step 1: Implement the technique for data centers and servers.
Step 2: Categorize the data centers and virtual machines.
Step 3: Obtain the maximum and minimum utilizations of resources.
Step 4: Check the pheromone persistence and again implement the technique.
Step 5: Obtain the fitness parameter and check whether it satisfies the threshold value and accordingly the tasks will be scheduled by allocating the virtual machines.

Step 6: Allocate the virtual machines using similar criteria.
Step 7: Evaluate the response time and processing time of each data center.

Particularly, this algorithm is associated with a single result based on the track used by ants for obtaining the solution. It may also be implemented to check the fitness value of the virtual machines to ensure that the virtual machine is fit and competent for the implementation of tasks. Accordingly, the data centers and virtual machines may be set with maximum threshold value of the resources (Figures 2.1 and 2.2).

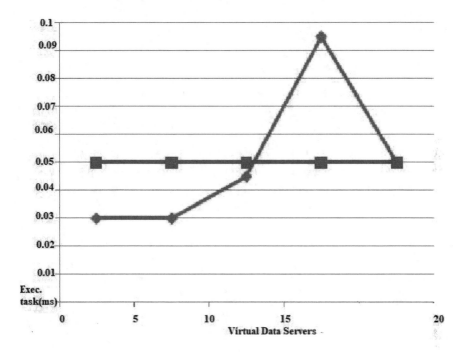

FIGURE 2.1 Virtual data servers with task execution.

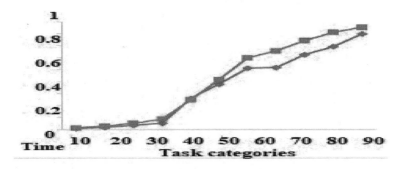

FIGURE 2.2 Task categorization and implementation.

Algorithm

Step 1: Obtain the pheromone value linked to each path between tasks and resources.

Step 2: Set the resources at random and fix the ants.

Step 3: Obtain the optimal or near-optimal solution iterating each ant.

Step 4: Obtain the fitness value through computation on each ant.

Step 5: Compare the optimal or near-optimal solution with the solution obtained through each ant with better fitness value, if its fitness value is better than optimal solution, change the optimal parameter.

Step 6: Update the local pheromone and global pheromone.

Step 7: Continue till the condition is achieved.

2.8   EXPERIMENTAL ANALYSIS

The simulation and performance analyses have been done considering the statistical metrics from the simulation linked to overall response time of the system and data server processing time along with execution time of tasks. The uniform result set modifies the dataset to obtain complete solution. It is also very essential to evaluate the fitness value of the virtual machines and to ensure that these are really efficient for implementation of tasks. After the scheduling process, the tasks may be mapped with the data centers to maintain equality while distribution of virtual machines and to improve the performance (Tables 2.1 and 2.2).

TABLE 2.1
Estimation of Virtual Servers

Serial Number	Virtual Data Servers	Execution Time of Tasks (ms)
1.	7	0.03
2.	9	0.039
3.	11	0.045
4.	15	0.05

TABLE 2.2
Implementation of Tasks in Virtual Machines

Serial Number	Number of Tasks Per Virtual Machine	Execution Time of Tasks (ms)
1.	40	0.40
2.	50	0.51
3.	60	0.61
4.	70	0.79

2.9 CONCLUSION

As task scheduling may be a major concern in cloud platform, it may be more essential to balance the load on virtual machines. In such scenario, the metaheuristic approach has been applied to obtain the near-optimal solution within relative time. The primary intention in this case is to improve overall response time, data center processing time, cost, and other performance metrics. Many times, it is essential to focus on greening data centers. To achieve the practical applications and measure the performance of data centers, it is necessary to estimate the overall cost of owner-ships and provide the minimum measuring concepts and techniques for greening the data centers. It is also necessary to focus on the strategies of accumulation of data in such situations.

REFERENCES

1. T. Ristenpart, Exploring information leakage in third-party compute clouds, presented at the *16th ACM Conference on Computer and Communications Security*, Chicago, IL, November 9–13, pp. 67–71, 2009.
2. F. Sabahi, Intrusion detection techniques performance in cloud environments. In *Proceedings of the Conference on Computer Design and Engineering*, Kuala Lumpur, Malaysia, pp. 398–402, 2011.
3. P. R. Gallagher, *A Guide to Understanding Data Remanence in Automated Information Systems*. The Rainbow Books, ch.3 & ch.4, 1991.
4. K. K. Fletcher, Cloud Security requirements analysis and security policy development using a high-order object-oriented modeling, M.S. thesis, Dept. Computer Science, Missouri Univ. of Science and Technology, Rolla, MS, 2010.
5. M. Kalra, and S. Singh, A review of metaheuristic scheduling techniques in cloud computing. *Egyptian Informatics Journal*, 2015, pp. 275–295, 2015.
6. M. A. Tawfeek, A. El-Sisi, A. Keshk, and F. Torkey, Cloud task scheduling based on ant colony optimization. IEEE 2014, Turkey, pp. 64–69.
7. R. Madhvi, and S. Sowmya Kamath, An hybrid bioinspired task scheduling algorithm in cloud environment. IEEE 2014, Suratkal, India, pp. 1–7.
8. B. Kruekaew, and W. Kimpan, Virtual machine scheduling management on cloud computing using artificial bee colony. In *International Multi Conference of Engineers and Computer Scientists*, Hong Kong, 2014.
9. K. Nishant, P. Sharma, V. Krishna, C. Gupta, K. P. Singh, Nitin, and R. Rastogi, Load balancing of nodes in cloud using ant colony optimization. In *International Conference on Modelling and Simulation*, Dehradun, India, pp. 20–27, 2012.
10. R. N. Calheiros, R. Ranjan, A. Beloglazov, C. A. F. Rose, and R. Buyya, CloudSim: A toolkit for modelling and simulation of cloud computing environment and evaluation of resource provisioning algorithms. *Software Practice and Experience*, 41, pp. 23–50, 2010.
11. S. Chen, J. Wu, and Z. Lu, A cloud computing resource scheduling policy based on genetic algorithm with multiple fitness. In *Computer and Information Technology(CIT)*, Chengdu, 2012.
12. X.-F. Liu, Z.-H. Zhan, K.-J. Du, and W.-N. Chen, Energy aware virtual machine placement scheduling in cloud computing based on ant colony optimization approach. In *Annual Conference on Genetic and Evolutionary Computation*, New York, USA, pp. 41–48, 2014.

13. J. T. Christina, K. Chandrasekaran, and C. Robin, A novel family genetic approach for virtual machine allocation. *Procedia Computer Science*, 46, pp. 558–565, 2015. Elsevier.

14. C. T. Lin. Comparative based analysis of scheduling algorithms for RM in cloud computing environment. *International Journal of Computer Science Engineering*, 1(1), pp. 17–23, 2013.

15. Z. Liu, W. Qu, W. Liu, Z. Li, and Y. Xu, Resource preprocessing and optimal task scheduling in cloud computing environments. *Concurrency and Computation Practice and Experience*, 27, pp. 3461–3482. Published online in Wiley Online Library (wileyonlinelibrary.com). doi:10.1002/cpe.3204, 2014.

16. J. Koomey, Growth in data center electricity use 2005 to 2010, A report by Analytical Press, completed at the request of The New York Times, 2011.

17. A. Banerjee, T. Mukherjee, G. Varsamopoulos, and S. K. S. Gupta, Integrating cooling awareness with thermal aware workload placement for HPC data centers. *Sustainable Computing: Informatics and Systems*, 1, pp. 134–150, 2011.

18. L. Wang, G. von Laszewski, J. Dayal, and T. R. Furlani, Thermal aware workload scheduling with backfilling for green data centers. In *Performance Computing and Communications Conference (IPCCC), 2009 IEEE 28th International*, Edinburgh pp. 289–296, 2009.

19. M. Maruyama, A. Takahashi, H. Yajima, A. Takeuchi, N. Yamashita, M. Matsumoto, and K. Tajima, Reducing electric power consumption for air conditioning by improving temperature distribution in telecom equipment rooms. *NTT Technical Review*, 11, p. 10, 2013.

3 A CMOS-Based Improved Wilson OP-AMP for Green Communication Applications

Suraj Pattanaik and Nirupama Deep
Gandhi Institute for Education & Technology
and
BPUT Rourkela

CONTENTS

3.1   INTRODUCTION

Green technology is a very advanced and new technology that has become popular due to its eco-friendliness. Presently, day by day our environment is becoming imbalanced, which has adverse effects upon us [9,10].

Using green technology, we can save our environment in the long term. Green technology implies development and application of products and systems that protect our environment as well as natural resources. The main feature of using this technology is that it reduces the negative impact of human activities on the environment. Reducing the use, recycling products, and renewing energy are the main goals of green technology. Green communication is a growing research field in wireless communication. The green communication plays an important role in maximizing efficiency while consuming low power in portable electronic devices [9].

In green communication or green technology, energy saving and energy consumption are the topics we need to consider now. Low power consumption equipment manufacturing is the challenging field in complementary metal oxide semiconductor (CMOS) technology [10]. With the rapid development in electronics field, it is now possible to design low-power and area-efficient very large-scale integrated digital circuits. Due to these characteristics, nowadays the electronic devices have become portable and user-friendly besides being less costly. So, now consumer can utilize digital electronic devices easily. Almost all people are using electronic devices or electronic gadgets every single day. So, to satisfy the people's need, the engineers or the designers and the developers have to deal with the parameters such as power consumption [11].

Operational amplifier (OP-AMP) is mostly used in analog integrated circuits, and it plays an important role in designing radio frequency (RF) applications using logic gates. CMOS technology–based OP-AMP consumes less power than traditional OP-AMP. For RF applications, high-speed, low-power, and high-gain OP-AMP is required, so the unity bandwidth (UGB) of OP-AMP is high because high-gain OP-AMP uses improved Wilson structure with long channel length transistor biased low current levels. High-bandwidth amplifier uses single bandwidth short channel length transistor in different techniques [2,3]. OP-AMPs can be used in different applications like summer, multiplier, subtractor integrator, active filter, differentiator, digital-to-analog converter, etc. Common mode is influenced by the mismatches and the load effect of the OP-AMP. Due to this phenomenon, it requires an extra common mode stabilization circuit which increases power consumption and complexity level of the circuit. To improve UGB, different techniques like doublet free pole-zero cancelation and Gm boosting techniques are used, but these techniques also consume extra power, so are not suitable for green communications and battery-operated devices [4,5]. The main disadvantage of bulk-driven CMOS technique is the latch-up effect due to which gain will be reduced. To improve the gain, positive feedback can be used but it also increases the complexity and power consumption. In traditional design, gain is high but UGB is not good. For green communication, RF applications require better matching of gain and phase margin with minimum power consumption. In CMOS technology, designing an OP-AMP with high UGB and moderate DC gain is a challenging task because of continuous decrease in channel length. To improve the gain of the device, we can

1. increase the number of stages,
2. increase the transconductance, or
3. increase the output resistance.

If the number of stages is increased, then stability of the circuit will be affected; if transconductance is increased, then drain current also increases initiating more power consumption.

In contrast, on increasing the output resistance, the gain increases and power consumption reduces. This chapter proposes a new approach on current mirror and improved Wilson technique–based OP-AMP for achieving low power consumption, high gain, and high slew rate [6,7].

3.2   GREEN COMMUNICATION

Communication technology plays an important role in our day-to-day lives, and a large amount of transmitting power is needed to support all communication devices. Green communication is the practice of selecting energy-efficient communication and networking technology which uses minimum resources. An increase in energy consumption results in excess of carbon dioxide emission. Thus, it is evident that a large amount of greenhouse gasses is produced by the cellular network devices, and excess of carbon dioxide emission caused by wireless networks affects human health, so a system having higher specific absorption rate (SAR) value will be harmful for both health and the environment. Being energy efficient is also important for extended battery life of wireless devices, that is, less energy consumption will result in reduced _carbon dioxide emission, therefore green communication is very necessary for human health as well as environment [12].

3.3   BANDWIDTH

Bandwidth is the most important concept in green communications. Shannon's capacity formula shows that the bandwidth is in direct relation to the transmission rate for a given amount of transmitted power. Energy consumption can be reusable with the bandwidth for a given data transmission rate [12].

3.4   OPERATIONAL AMPLIFIER

OP-AMP is a high-gain electronic voltage amplifier with a differential input and, usually, a single-ended output. To make a high-speed OP-AMP, its UGB should be very high. Since most of the portable devices are battery operated, the power consumption of the designed circuit should be low [1].

3.5   CASCODING IN SECOND STAGE OP-AMP

In this circuit design, a current mirror-based cascade stage is used in the second stage as shown in Figure 3.1. Here the transistor M1 and M2 are the input differential pair. M3 and M4 are the active loads. In the output stage, M6 and M7 form the positive channel metal oxide semiconductor (PMOS) cascade stage. Here, M6 is the common source transistor which drives the common gate M7 transistor. M11 and M7 make a PMOS current mirror. The current through M7 will be a fraction of

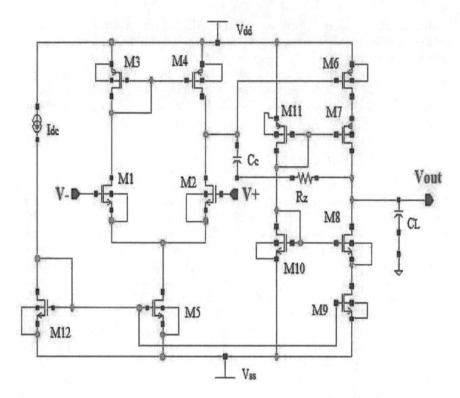

FIGURE 3.1 CMOS OP-AMP with current mirror-based cascoding at the second stage [1].

current through M11 and this fraction is determined by the ratio of aspect ratio of M7 to the aspect ratio of M11. M8 and M9 transistors form the n-channel metal oxide semiconductor (NMOS) cascade stage. M8 and M10 make a NMOS current mirror. The current through M8 is based on the aspect ratio of M10. M12 transistor makes a current mirror configuration with M5 transistor while providing DC bias to the gate of the M5 transistor [1].

In Figure 3.1, M9 is the common source transistor, which drives the common gate M8 transistor. Here M11 transistor makes a current mirror with transistor M7 and supplies DC biasing to the gate of M7 transistor. M10 transistor makes a current mirror with M8 transistor and supplies DC biasing to the gate of M8 [1,2]. Considering the small signal analysis of Figure 3.1, we found that the small signal voltage gain of this circuit is given as

$$A_v = g_{mI} g_{mII} R_I R_{II}$$ (3.1)

The transit frequency of a MOS transistor is given by the following equation:

$$f_t = \frac{g_m}{2\pi \left(C_{gs} + C_{gd} \right)}$$ (3.2)

where

$$C_{gs} = \frac{1}{2} WLC_{ox} + WL_{ov}C_{ox} \left(\text{Triode region} \right) \tag{3.3}$$

$$= \frac{2}{3} WLC_{ox} + WL_{ox}C_{ox} \left(\text{Saturation region} \right) \tag{3.4}$$

$$WL_{ox}C_{ox} \left(\text{Cut off region} \right) \tag{3.5}$$

and

$$C_{gd} = WLC_{ox} + WL_{ov}C_{ox} \left(\text{Triode region} \right) \tag{3.6}$$

$$= WL_{ov}C_{ox} \left(\text{Saturation region} \right) \tag{3.7}$$

$$= WL_{ov}C_{ox} \left(\text{Cut off region} \right) \tag{3.8}$$

From Equation (3.1), we found that the transit frequency is inversely proportional to the channel length. In the proposed circuit, channel length of 45 nm is used, which is very small. So it produces a very high UGB. Equations (3.3) and (3.6) represent the values of C_{gs} and C_{gd} at different regions such as triode, saturation, and cut-off [1,2].

3.6 PROPOSED WORK

3.6.1 Current Mirror-Based Improved Wilson at Second Stage

In this circuit, the transistor can operate at very low power supply which leads to low power consumption by the transistor. In the proposed circuit (design 1), the two transistors count is increased, so the overall power consumption is very less and gain is also increased.

In this circuit, a current mirror-based improved Wilson stage is used in the second stage as shown in Figure 3.2a. Here transistors M7, M11, M13 and M8, M10, M14 also form the improved Wilson current mirror where M6 and M9 are the common source transistors.

3.6.1.1 Design 1

The proposed CMOS OP-AMP with improved Wilson current mirror is designed and tested using Cadence Virtuoso version (6.1.6–64b.500.4). Figure 3.2a shows the schematic design, and Figure 3.2b shows the transient response of input and output.

3.6.1.2 Design 2

In this circuit, the transistor can operate at very low power supply which leads to low power consumption by the transistor. In the proposed circuit (design 2), the two number of transistors count is increased in the second stage of the OP-AMP. So, the gain of the circuit is increased, and the power consumption is minimized because 45 nm technology is used.

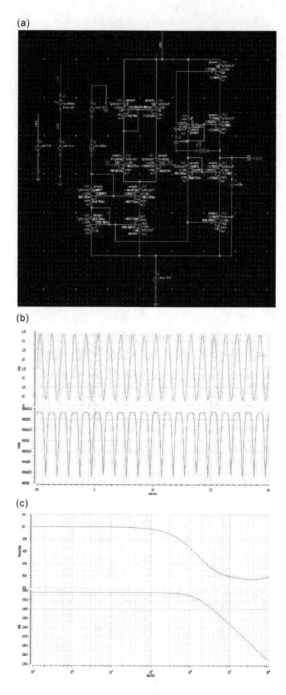

FIGURE 3.2 (a) Schematic of CMOS OP-AMP with improved Wilson current mirror design 1. (b) Transient response of CMOS OP-AMP with improved Wilson current mirror design 1 input and output. (c) CMOS OP-AMP with improved Wilson current mirror design 1 phase and gain versus frequency curve.

3.6.1.3 Design 3

In this circuit, the transistor can operate at very low power supply which leads to low power consumption by the transistor. In the proposed circuit (design 3), overall the four number of transistors count is increased. In the second stage of the OP-AMP, 10 transistors are used. So, the gain of the circuit is increased and the power consumption minimized.

3.7 RESULTS AND DISCUSSION

The proposed OP-AMP circuits have been designed and simulated in 45 nm CMOS process by Cadence analog and digital tools. The simulation result of proposed OP-AMP (design 1, design 2, and design 3) is compared with cascoding in second stage current mirror OP-AMP. AC response of proposed OP-AMP with same capacitor load. The CMOS OP-AMP with cascading in second stage design 1, design 2, and design 3 produce, respectively, an UGB of 1.9, 2.4, and 3.7 GHz; a DC gain of 119, 142, and 150 dB; a phase margin of 145°, 108°, and 130°; a slew rate of 1,231, 1,638, and 2,131 V/μs; and a power dissipation of 0.33, 0.56, and 0.55 mW. The proposed OP-AMPs are given in Figures 3.2a, 3.3a, and 3.4a. The UGB of proposed OP-AMP design 3 is greater than other designs. As demonstrated in Figure 3.4a, the proposed OP-AMP achieves a 150 dB gain, which is greater than other designed OP-AMPs.

When an OP-AMP is used for practical application, the operating temperature and power supply voltage are in the normal range. Three types of current mirrors are implemented in the OP-AMP, and the results are compared. The results are tabulated in Table 3.1.

3.8 POWER MEASUREMENT OF THE PROPOSED WORK

The average power calculation is done in Cadence Virtuoso ADE L calculator window and measured for different frequencies. The power consumption values of the proposed improved Wilson current mirror design 1, design 2, and design 3 are 0.33, 0.56, and 0.55 mW, respectively.

3.9 CONCLUSION

In this chapter, three types of new OP-AMP design techniques have been proposed, which use a current mirror-based improved Wilson at the second stage of an OP-AMP. The proposed techniques increase the gain bandwidth product of the OP-AMP. So, the improved Wilson technique produces better results than other cascoding techniques using gain and phase plots. The main advantage of the circuit is that it produces an overall very good result for wireless applications while consuming low power, which makes it suitable for green communication. The circuit is designed as highly linear circuit. When compared with previously designed circuits, superior gain, bandwidth, slew rate, and power dissipation have been observed in the circuit.

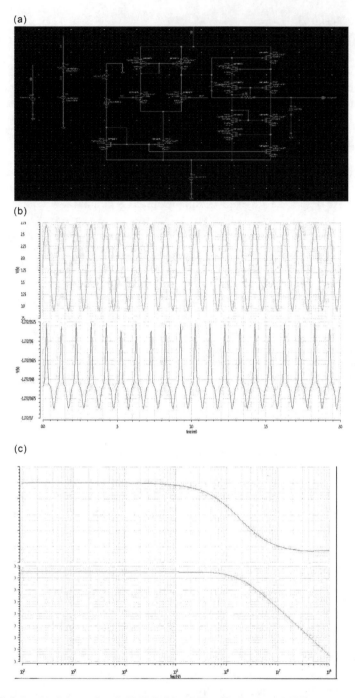

FIGURE 3.3 (a) Schematic of CMOS OP-AMP with improved Wilson current mirror design 2. (b) Transient response of CMOS OP-AMP with improved Wilson current mirror design 2 input and output. (c) CMOS OP-AMP with improved Wilson current mirror design 2 phase and gain versus frequency curve.

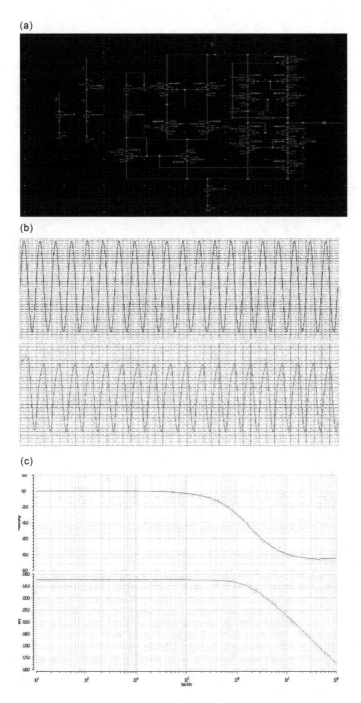

FIGURE 3.4 (a) Schematic of CMOS OP-AMP with improved Wilson current mirror design 3. (b) Transient response of CMOS OP-AMP with improved Wilson current mirror design 3 input and output. (c) CMOS OP-AMP with improved Wilson current mirror design 3 phase and gain versus frequency curve.

TABLE 3.1
Performance Comparison Table of Different OP-AMP Designs

Parameter	Cascode Current Mirror [1]	Proposed Improved Wilson Current Mirror (Design 1)	Proposed Improved Wilson Current Mirror (Design 2)	Proposed Improved Wilson Current Mirror (Design 3)
Gain (dB)	63.62	119	142	150
Power Dissipation (mW)	0.71	0.33	0.56	0.55
Phase Margin (°)	59.53	145	108	130
UGB (GHz)	2.7	1.9	2.4	3.7
Slew rate (V/μs)	1816	1231	1638	2132
C_l (pF)	10	10	10	10
V_{DC} (V)	1.8	1.8	1.8	1.8
CMRR (dB)	6.87	29.33	43.62	41

REFERENCES

1. R. Kundu, A. Pandey, S. Chakraborty, V. Nath. A Current mirror based two stage CMOS cascode OP-AMP for high frequency application. *Journal of Engineering Science and Technology*, Vol. 12, No. 3 (2017), 686–700.
2. S. Bandyopadhya, D. Mukharjee, R. Chatterjee. Design of two stage CMOS operational amplifier in 180 nm technology with low power' and high CMRR. *International Journal of Recent Trends in Engineering and Technology*, Vol. 11, (2014), 239–247.
3. P. Ranjan, N. Dhaka, I. Pant. Ashutosh pranav design and analysis of two stage op-amp for bio-medical application. *International Journal of Wearable Device*, Vol. 3, No. 1 (2016), 9–16.
4. H. Langalia, S. Lad, M. Lolge, S. Rathod. Analysis of two-stage CMOS OP-AMP for single-event transients. *International Conference on Communication Information and Computing Technology (ICCICT)*, Oct. 19–20, Mumbai, 2012.
5. C. L. Kavyashree, M. Hemambika, K. Dharani, A. V. Naik, M. P. Sunil. Design and Implementation of two stage op-amplifier using 90NM technology (ICISC–2017), IEEE Explore, South Korea, 2017.
6. C. Zhang, A. Srivastava, P. K. Ajmera. A 0.8 V ultra –low power CMOS operational amplifier design. Department of Electrical and Computer Engineering, IEEE Explore, South Korea, 2002.
7. K. Saktidasan Sankaran, K. E. Purushothaman. Adaptive enhancement of low Noise amplifier using Cadence Virtuoso tool. IEEE Member Adhiparasakti Engineering College, Melmaruvathur, 2017.
8. R. W. Blaine, student Member IEEE, N. M. Atkinson, student Member, IEEE, J. S. Kauppila, Member, IEEE, T. D. Loveless, Member, IEEE, S. E. Armstrong, Student member, IEEE, and Lloyd W. Massengill, Fellow, IEEE. IEEE Transaction on Nuclear Science August, 2012.
9. A. M. Abdelbar, A. M. Ei-Tager, H. S. El-Hennawy. Radio frequency power amplifier for green communication, IEEE, NRSC, 2018.

10. W.-S. Chu, D.-M. Chun, S.-H. Ahn. Research advancement of green technologies. *IJPEM*, Vol. 15, No. 6 (2014), 973–977.
11. B. Khan, S. Pattanaik, Design A 4 –Bit carry look ahead adder using pass transistor for less power consumption and maximization of speed. In: Borah, S., Emilia Balas, V., Polkowski, Z. (eds.) *Advances in Data Science and Management. Lecture Notes on Data Engineering and Communications Technologies*, vol. 37. Springer, Singapore, 2020. ISBN-978-981-15-0977-3.
12. N. A. Malik, M. Ur-Rehman, Green communication: techniques and challenges. *EAI Endorsed Transaction on Energy Web and information Technology*, Vol. 7 (2017), e4.

4 Design for Manufacturing Automotive Components
A Knowledge-Based Integrated Approach

Sudhanshu Bhushan Panda, Antaryami Mishra, and Narayan Chandra Nayak
Indira Gandhi Institute of Technology

CONTENTS

4.1 INTRODUCTION

The automotive sector provides a vast market for industry verticals those elaborate in designing, manufacturing, and evolution of products, marketing, and selling of motor vehicles. The most challenging issue is to produce quality components that must meet all the application conditions, global standards, and are best in price and time [1,2]. This could be achieved by number of facts and figures: one of these is the best part design for the particular component with respect to its applications, standards, and acceptable unit cost. And the second most significant factor is to have reliable tool design that anticipated for defect-free part during injection molding production. The expanding markets for plastic materials and its application into today's competitive fast-growing areas such as consumer electronics, automotive, and appliances compel to produce parts with emerging quality, cost, and in limited time. That eagerly looking for alternatives to produce flawless parts, it is required to forecast the product quality before investing in manufacturing in terms of cost and time, which is the call at present to compete in market. The quality of injection molded components at a large depends on the processing variables of the process. It is complicated and inclusive of many process characteristics, for instance, temperature, time, and pressure. Modifying any of such variables will affect one or all, as they are interrelated with each other. To maximize the production capacity and quality of parts, the processing parameters need to be optimized. Designers/engineers are assisted by advanced software tools for injection mold flow simulations to visualize the problem areas and decide suitable injection molding process values. So, depending on observation, unreasonable or empirical method can be eliminated [3,4]. The most familiar process used to produce polymer components is the injection molding technique. It is one of the principal commercial activities for producing complex, net-shaped, three-dimensional (3D) plastic components [5]. In general, the technique of injection process covers polymer part planning, tool planning or design, and injection process parameter characterization, those all playing a major role in delivering the quality product and efficient production. Molding process of injection technique is generally segregated into five basic stages: melting, stuffing, packing, cooling, and ejection [4,6–8]. Mold cavities are required to be filled in a uniform time so as to get quality part [9]. Injection molding involves numerous parameters for part and tool design that need to be considered in an integrated and concurrent approach. It is also equally important to have flexibility in designing solid models for parts and molds by simultaneously incorporating design intent and optimization results. This system of design covers modules for calculation and selection of mold components. It makes use of advanced modern techniques associated with computer-aided design (CAD) software and simulation results as inputs for selection, modification, and design. It has been also analyzed by computer-aided design and engineering (CAD/CAE) integration [4,10–12]. The geometric features and design intent of plastic parts are also investigated through CAD/CAE for related analysis. Product geometry, in general, includes the information of a plastic product regarding its section thickness and mounting attributes such as ribs, bosses, guesses, chamfers, and holes; and the engineering attributes comprise interpretation-related information on design and sub-wall/evolve elements. It is also seen as a CAE structure for tool planning as well as manufacturing and variable computation for injection molding process [11]. Morphology matrix and decision diagrams are applied to analyze the process. All these processes are cast-off for rheological, mechanical and thermal computations, and raw material particulars and its details, but there it is not associated with computer-aided software tool [4,13]. Based on parametric and feature recognition, advanced software tools were used for 3D part modeling and mold base creation for plastic injection molding design system. This software provides a communicative platform to assist designers in quick creation of part modeling and tool planning and foster to make it a standardized process of mold design and planning. The software has options for calculations for number of impressions, mold components and dimensional verifications, and choice of injection molding machine. The orthodox unproved practice is costly, time-consuming, and insignificant. To make better tool plan, process controlling of injection molding variables and getting larger credence in CAD and CAE is required [6]. The tool design structure was developed by employing an unbarred application program interface (API) and mercantile computer-aided design, manufacturing and engineering (CAD/CAM/CAE) solutions [10,12,14]. Product designers were able to optimize part shape and sizes, and, at the same time, tool designers could optimize mold design systems. CAE principle is used to compute mold design and injection molding parameters with the help of numerical simulations. This is possible because of countless studies that were done on CAE software to study mold and characterization of injection molding processes. Regardless, each one of the earlier proposed procedure is not capable of forecasting the tool design and polymer injection process characteristics. This leads to draw the need of developing a software structure with unified process attributes and which can verify with the consequences based on computation and simulation of polymer molding, tool verification, and choice alternatives. Each of the research studies can be combined together for CAD and CAE integration of tool design process for polymer components [4,12]. For a prolonged time, number of researchers concentrated on tool designing of polymer injection molded parts facilitated with computers and software tools. Several researchers have made headway to series of measures that facilitated professionals and engineers to design components, tools, and deciding factors for characteristics of polymer molding in injection process. In the meantime, researchers also explored the standardization and integration of plastic injection mold design with CAD and engineering system of design, development, calculations, and modifications [4]. Many researchers enlightened mold flow simulation to optimize injection molding parameters only, but no research has been done to optimize the plastic part design and tool design as an associative integrated technique from CAD software and mold flow simulation results [4].

4.1.1 THE INVESTIGATED PART

In this research, an injection molded part is analyzed for flawless production from an original automotive vehicle manufacturer [3]. There are numerous visual and functional defects with the investigated part, such as sink marks appear on class A surface, which is not acceptable. Further, the flashes, deep machining marks, and other manufacturing scratches and defects should be positively avoided without changing the class A surface. There is a scope to investigate part design and tool design to eliminate all the visible defects or distortions during injection molding [10]. The injection molding process parameters also need to be taken into account to improve the processing defects [1,15]. Mold flow simulation helps designers to predict product designs without actual injection molding processing that saves time and cost [16].

4.1.2 OBJECTIVES

The specific purpose of this investigation is aimed at

- i. predicting problem areas of mold flow plastic part advisor;
- ii. determining the design experiments (DOE) and iterations of geometrical shape and size of the part to get optimized design.

4.2 MATERIALS AND METHODOLOGY

The plastic raw material used for this experimental analysis was A3WG6 (Ultramid)-BK00564, which is the reinforcement of 30% glass fiber; black pigment, which enhances heat and aging hostility to PA66-injection molding grade; housings and components for machines, which have high stiffness and dimensional stability. This engineering polymer is suitable for automobile component applications in harsh environmental conditions. This selected polymer has distinctive industrial uses, for example, for making vessels for automobile cooling systems, encase silhouette for window frames of aluminum, in cooling fans, and for making housings for light sockets [17]. The experiment was conducted using the 3D CAD data from the existing plastic part. The CAD data were acquired using Pro/ENGINEER software, which is commercially available for modeling and designing a variety of products [13]. The experimental part is optimized on geometrical dimensions, and the 3D CAD model is developed with help of the software. DOE has been done on part geometrical shape and size of the CAD model and is analyzed and optimized through mold flow plastic part advisor, an integrated simulation tool within Pro/ENGINEER software. The suggested methodology blends this Pro/ENGINEER tool with an especially evolved factor for the computation of polymer molding values, tool planning, and preference of tool components. This process of coalition employs invariable and CAD/CAE attributes to establish organized information to smoothen the procedure of reviewing, editing, and designing [18]. Generally, the output results and part structure drawn out of the CAD/CAE unified injection molding design structure are quite efficient and reliable [4,13]. Nevertheless, putting into practice these simulation results helps in the development of processing variables to acquire an empirical procedure. In reality, ascertaining the enhanced process characteristics for manufacturing an improved version of standard injection molding polymer component, the simulation experiment should fetch as many iterations. Accordingly, this investigational course of actions is unmanageable and disordered [6]. As a consequence, to overcome these circumstances, improved procedures are commonly combined for simulation so as to assist professionals in reaching out optimum processing parameters for framing up and establishing injection molding process [3,4,6].

4.3 EXPERIMENTS AND OPTIMIZATION

Tool design is a detailed congenial process that includes parametric, compatible, and feature-based integrated CAD/CAE system. Numerical simulation plays a major role and unlocks novel opportunities of part analysis in the course of the product and tool design. The part designing is analyzed and optimized by considering foremost process variables of injection molding for the specific plastic raw material as specified by the raw material manufacturer, BASF Corporation [17]. Further, by taking the results from the analysis as input, the part design get an optimized number of times till satisfactory results are obtained, and finally, the best optimized part design is anticipated. The motivation behind presiding over the evaluation is to distinguish and adopt the correlated molding characteristics that influence mechanical values of the molded product. Within this research, mold flow plastic advisor analysis and calculations are used to perform all simulations. For the mold flow analysis, overall the part dimensions were taken as 225 mm length×39 mm width×44 mm height. This investigation considers the meshing model for simulation with 200,599 number of elements in part and 171,814 number of nodes. This research integrates 3D designing software tools with mold flow simulation tool and for validation it is based on the results of part geometry and design. The existing part is analyzed visually to notify all the possible defects and systematically categorized strenuously. The CAD data were imported to mold flow simulation platform.

All the injection molding process input data to carry out the analysis are considered to hold the raw material manufacturer's data information sheet for the material: Ultramid A3WG6–PA66–GF 30, 180 MPa as injection pressure, 290°C as melt temperature, 85°C as mold temperature, and the density of the part is taken as 1.36 g/cm³ [17].

Material selection and conformity to recognized standards of data play a major role in delivering the best results [1]. A balanced understanding is required to match the process and material parameters. The various stages of injection molding part design, to avoid quality issues, are depicted in a logical flow diagram, as shown in Figure 4.1. Most of the injection molding components of the complex, 3D structural profiles, and the rheological characteristics of the molten plastic do not adhere to Newtonian and isothermal laws. With these multiple unprecedented problems in computational analyses, mainly during the process of filling, precautions were required to be taken [19].

The injection molding variables, namely, the temperature of polymer melt, pressure of injection, pressure of holding, mold temperature, and filling duration, are important parameters to consider during molding process [13]. The mold flow virtually has been done in the experiments to foresee the problem areas in the particular part, and subsequently inputs are taken to modify the part geometry and design.

A number of iterations were done on part design till the best results were achieved from the simulation experiment. This activity was repeated till satisfactory results in terms of part design and molding process parameter values were acquired.

4.3.1 Analysis of the Part

The 3D parametric CAD data were fetched from Pro/ENGINEER software and analyzed using Mold flow Plastic Part Advisor for preferable part design and to predict injection molding process parameters. Following observations were noted after multiple workouts of the product.

More sink marks are concentrated at the top surface, as shown in Figure 4.2a, blue-shaded color gradually changes to green, yellow, orange, and finally to red color shades indicating severe shrinkage marks on functional class A face, which could be visible with the tangible product and in connection with the result of simulation published. In Figure 4.2b, the filling time of cavities is indicated in seconds. Figure 4.2c reflects non-uniformity pressure for injection dispersal in the course of injection processing (red to blue tone shading) while utmost pressure for injection is equitable. In Figure 4.2d, drop in pressure at one end is visible from red tone, which is also required to be fairly in proportion. The area in color yellow reflects inferior

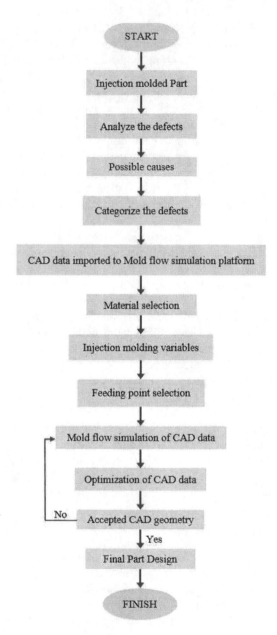

FIGURE 4.1   Flow chart of part defect analysis.

standard of the part, which could be due to increased wall thickness as compared with other sections of the component, as illustrated in Figure 4.2e. According to the simulation results, the polymer weld marks are liable to rupture [20] at mentioned specific location of metallic part as clearly visible opposite to ribs as well as feeding point (shown in Figure 4.2f), which states the need of reducing the marks as

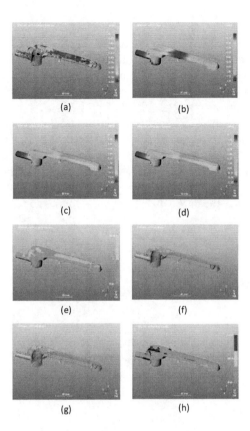

FIGURE 4.2 Mold flow analysis for existing part. (a) Shrinkage area. (b) Fill time. (c) Injection pressure. (d) Pressure drop. (e) Quality prediction. (f) Weld lines. (g) Air traps. (h) Cooling quality.

much as feasible. In Figure 4.2g, the gas-trap is shown, which is the foremost vital feature for injection part value and class. This necessitates the part to leave the mold cavity in course of packing stage to get rid of bubbles in the component, which in aftermath degrades the part strength and introduces defective components [2]. Figure 4.2h points out the cooling status of the part toned with red shades, which implies indigent part section geometry and thickness at those locations that have to be eliminated by keeping invariable part thickness, and, if possible, to lessen the thickness particularly on affected regions.

4.3.2 PART DESIGN OPTIMIZATION

After a brief design of geometrical shape and size iterations, it is found that the most favorable part design is represented as follows:

For the sake of the refinement of the product, the results of analysis data should be linked with the drawbacks of the part. The specific reasons were studied and the end result is listed here for the enhancement of the component.

1. Design of ribs

   In Figure 4.3a and b, it is shown that the part has irregular wall thickness and bulky section at top, which creates more sink mark on class A surface. But the maximum rib thickness should be at $t = 0.75$ T, where "T" is the general section thickness of the product that should be taken to avoid sink mark at rib section area [21,13].

   For better control of sink marks, ribs should be thin, as shown in Figure 4.4a and b, compared with existing design and the gap between the ribs be increased to minimize the sink mark and to retain the required strength. All the edges and corners should be filleted for better plastic flow and minimal stress development, as the rib design affects the part strength and durability of the part with respect to its application [21].

2. Insert molding

   The metal insert outer wall is not uniform, as shown in Figure 4.5a and b, for which sink marks are heavily seen just opposite to its placement, on top surface. According to the simulation results, shown in Figure 4.2a, sink marks are accumulated in this area.

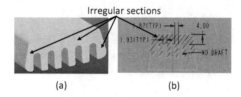

(a)   (b)

FIGURE 4.3   Rib design with existing part.

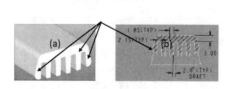

FIGURE 4.4   Rib design with optimized part.

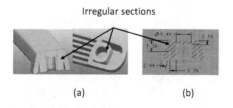

(a)   (b)

FIGURE 4.5   Metal insert design with existing part.

Uniform sections

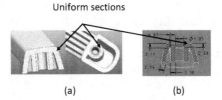

(a) (b)

FIGURE 4.6 Rib and metal insert design with optimized part.

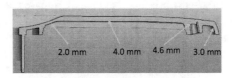

2.0 mm   4.0 mm   4.6 mm   3.0 mm

FIGURE 4.7 Existing part with variable wall thickness.

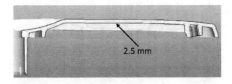

2.5 mm

FIGURE 4.8 Optimized part with uniform wall thickness.

The metal insert with respect to the top wall thickness and around it is maintained uniform and kept of minimum possible thickness so as to avoid sink marks during molding [13], as shown in Figure 4.6a and b. To keep minimum wall thickness for better control of sink marks, the bulky gathering of polymers is avoided, as shown with the existing part shown in Figure 4.5.

3. Wall thickness

The overall wall thickness variations on the part (Figure 4.7), being the crucial obstacle to the plastic melt flow together with correlative composition of component at discrete sections, create the very large shrinkage spots and also impair the durability and standard grade of the product [22–24]. The improved product design has prevailed over those all setbacks by maintaining uniform section thickness on these locations as shown in Figure 4.8.

4.4 RESULTS AND DISCUSSION

Brief results are obtained ofrom the "Mold Flow Plastic Product Advisor" analysis by running experiments repeatedly for a number of times on the software with the inputs for the specified raw material to get the new optimized part design.

The geometrical shape and size of the part is modified to get best possible outcome based on injection molding parameters as per the specification of particular plastic raw material. Once the CAD data are finalized, the part is analyzed through mega pascal (MPA), and after repeated number of cycles the best results are shown in Figure 4.9a–h.

The sink marks, as shown in Figure 4.9a, are minimized as compared with the existing part. The higher value of filling duration, as given in Figure 4.9b, regulates the concerned part's quality [3]. The pressure for injection is almost balanced as shown in Figure 4.9c, and it signifies that the feeding point is at right location. Pressure drop, as shown in Figure 4.9 d, is related to the part section thickness and volume of the plastic part with respect to the feeding point. The analysis results for overall quality of the part are also better as compared with the existing part (Figure 4.9e). Weld lines, as shown in Figure 4.9f, inside the polymer deteriorate the toughness and durability of the component in real applications [1]. Weld zone is generally formed at the flow front of polymers inside the mold, where two or more directional polymer melts get converge. Air traps or voids, as shown in Figure 4.9g, appear inside the part during injection of molten polymer due to hindrance of gases from the mold. The comprehensive cooling quality [15,25], as shown in Figure 4.9h, implies the cooling provisions for cooling passages for cavities and cores, particularly while designing tool. While establishing the simulation outcomes, the values of melt temperature of polymer, pressure of packing, and duration for injection come were found significant [3,26].

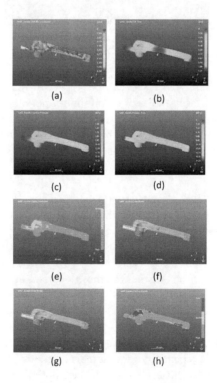

(a)

(b)

(c)

(d)

(e)

(f)

(g)

(h)

FIGURE 4.9 MPA analysis of optimized part. (a) Sink area. (b) Fill time. (c) Injection pressure. (d) Pressure drop. (e) Quality prediction. (f) Weld lines. (g) Air traps. (h) Cooling quality.

The sink marks, as shown in Figure 4.9a, are reduced compared with the existing part design, and it shows the design and wall thickness have played an important role on sinking ability of the plastic parts. The analysis also helps to find out the short volume which was 43.4 cm³ and is reduced to 40.7 cm³, which is 6.22% less, as shown in Table 4.1; the filling time is 1.21 s, which is 8.33% less to the existing part; whereas the suitable injection pressure is calculated at 17.2 MPa, which is around 28% more than the existing. The simulation of part is also helpful for better packing of polymer inside the mold to get better mechanical properties of the molded component [11]. The clamping force is also increased to 11.2%, which is helpful for the part to be flash free [7,12,14,25]. The most important injection molding defect is due to sink of the polymer after molding [22]. Warpage is a prime factor that affects product quality [27]. Non-uniformity of cooling, shrinkage, and orientation are heavily influenced warpage [28]. The result here shows 58.62% of sink ability, which is improved with the new optimized part design, and more specifically class A surface has a minimum number of sink marks. The quality prediction is 37.8% better as compared with the existing part. The analysis result also indicates the cooling quality, as shown in Table 4.1, which has improved to 87%, which is 25.3% more than the existing part. Depending upon the gate location and the part design, the polymer flow front and joining are also improved [13] by virtue of which the weld quality of the plastic part gets improved.

In this research, the part design, injection molding process parameters, etc. are well balanced to get the best results regarding weld lines. The weld lines are decreased to 58.4% as compared with the existing part. Earlier, the air trap inside the part during injection molding was also a significant defect to the quality of the

TABLE 4.1
Comparison of MPA Results

Parameters	Existing Part	Optimized Part	Remarks
Raw material	Ultramid A3WG6	Ultramid A3WG6	
Maximum injection pressure (MPa)	180	180	
Mold temperature (°C)	85	85	
Melt temperature (°C)	290	290	
Shot volume (cm³)	43.4	40.7	6.22% decrease
Fill time (s)	1.32	1.21	8.33% decrease
Injection pressure (MPa)	12.4	17.2	27.9% increase
Filling clamp force (tonnage)	4.44	5.0	11.2% increase
Pressure drop	Average	Better and balanced	
Sinkability (%)	29	12	17% decrease
Sink marks (%)	High on Class "A" surface	Low on Class "A" surface	
Quality prediction (%)	56	90	34% increase
Cooling quality (%)	65	87	22% increase
Weld lines (%)	12	5	7% decrease
Air traps (%)	30	20	10% decrease

part [12,25], and now it is also minimized, that is, it is reduced to 33.33% as compared with earlier part design. So, finally the new optimized injection molded part ensures better quality as compared with the existing design. The gate diameter or feeding opening plays an important role on quality of the injection molded part [7], particularly the sink marks. It is noticed the gate diameter of 3 mm in the existing part incorporated more sink marks, and when the gate diameter was reduced to 2 mm, the sink marks also reduced considerably. Unlike gate diameter, gate location has a considerable effect on the quality of injection molded part [29]. The gate location versus the percentage of sinkability is analyzed and the best gate location was found to be as 62 mm in x-direction and 6 mm in y-direction. The minimum possible sinkability is shown to be 22%, compared with other gate locations. Overall quality of the part is also analyzed by iteration of gate diameter and it is found that the best suited gate diameter for this particular part is 2 mm, and the best quality of the part is predicted at 85%. The gate location versus overall quality of the part has also been analyzed and the best quality of the part was found to be 85% with the gate locations finalized at 62 mm in x-direction and 6 mm in y-direction.

It is found that mold surface temperature and melt temperature are the prime process control parameters for warpage and sink marks [12,25] in injection molding of plastics. The factors influencing the quality of a molding component are design of part, design of mold, performance of machine, and conditions of processing [14]. Moldflow Insight Plastics software helps to investigate the factors affecting polymer part and tool designing during the time of injection molding process, for example, short impression, dissimilar filling, over packing, weld areas, etc. [8,26]. It also helps to keep away from bulky geometries by preserving similar section thickness to minimize the price of raw materials and as well machine time [20]. Mold simulation software contributes adequate statistics in respect of its duration for filling [2,9], pressure concerning injection, and pressure decreases [16], which could abstain from the defects in polymers, made from factual injection molding, like shrinkage points, air bubbles, and extra packing [18]. Many of these crucial decisions should be taken in the design phase of complex injection molding [27], which saves total costs and product timeline [22]. It is also concluded that by controlling the process parameters, the depth of a sink mark can be minimized or completely eliminated [23,24]. Implementation of CAD/CAE and CAM during manufacturing and design of tools for injection molding for the automobile parts followed by simulation [1,13,30] furnishes tool design engineers and concerned manufacturers with a considerate way of polymer behavior during the course of its solidification for the metallic insert component. So they are now able to correctly design and effectively use polymer solidification techniques. The optimized model could deliver correct outcomes by simulation results.

4.5 CONCLUSIONS

A virtual reality–based product designing method for plastic part design is developed that could save cost, time, man-hour, energy in this competitive and fast changing world. The times for filling and packing are closely related to each other, so they affect the sink marks and dimensional quality of the plastic part. The experimental work was much easier and time saving. This simulation mechanism in association with advanced software tools motivates engineers to assure their product design's credence in production to compete into market. The proposed system also saves man-hour, resources, and minimizes efforts to produce injection molded parts. This virtual-reality system of techniques involves artificial intelligence principles before actual part production facilitates to compete in the process of plastic product development in today's fast-changing technological markets. To conceive a sustainable product, it may carefully be exploited in designing the manufacturing methodology.

REFERENCES

1. Mayyas, A., Qattawi, A., Omar, M., & Shan, D. (2012), Design for sustainability in automotive industry: A comprehensive review. *Renewable and Sustainable Energy Reviews*, 16(4), 1845–1862. doi:10.1016/j.rser.2012.01.012.
2. Katarína Szeteiova, I. (2010), Automotive materials: Plastics in automotive markets today, 27–33.
3. Mehat, N. M. & Kamaruddin, S. (2011), Investigating the effects of injection molding parameters on the mechanical properties of recycled plastic parts using the taguchi method. *Materials and Manufacturing Processes*, 26(2), 202–209, doi:10.1080/104269 14.2010.529587.
4. Matin, I., Hadzistevic, M., Hodolic, J., Vukelic, D., & Lukic, D. (2012), A CAD/CAE-integrated injection mold design system for plastic products. *The International Journal of Advanced Manufacturing Technology*, 63(5–8), 595–607. doi:10.1007/s00170-012-3926-5.
5. Guo, W., Hua, L., Mao, H., & Meng, Z. (2012), Prediction of warpage in plastic injection molding based on design of experiments. *Journal of Mechanical Science and Technology*, 26(4), 1133–1139. doi:10.1007/s12206-012-0214-0.
6. Zhou, H., Shi, S., & Ma, B. (2009), A virtual injection molding system based on numerical simulation. *The International Journal of Advanced Manufacturing Technology*, 40(3–4), 297–306. doi:10.1007/s00170-007-1332-1.
7. Elsheikhi, S. A., & Benyounis, K. Y. (2016), Review of recent developments in injection molding process for polymeric materials. Reference Module in Materials Science and Materials Engineering. doi:10.1016/b978-0-12-803581-8.04022-4.
8. Chen, P.-H. A., Villarreal-Marroquín, M. G., Dean, A. M., Santner, T. J., Mulyana, R., & Castro, J. M. (2018), Sequential design of an injection molding process using a calibrated predictor. *Journal of Quality Technology*, 50(3), 309–326, doi:10.1080/0022406 5.2018.1474696.
9. Osswald, T. A., & Hernández-Ortiz, J. P. (2006), *Polymer Processing - Modeling and Simulation*, Hanser Gardner Publications, Cincinnati.
10. Nee, Andrew Y. C. (2003), Computer-aided tooling design for manufacturing processes. *Annals of the CIRP*, 46(3), 429–432.
11. Lee, S. H. (2009), Feature-based non-manifold modeling system to integrate design and analysis of injection molding products. *Journal of Mechanical Science and Technology*, 23(5), 1331–1341. doi:10.1007/s12206-009-0407-3.
12. Kohlhase, M., Lemburg, J., Schröder, L., & Schulz, E. (2010), Formal management of CAD/CAM processes. *German Federal Ministry of Education and Research*, 28(5), 16–32.
13. Grujicic, M., Sellappan, V., Pandurangan, B., Li, G., Vahidi, A., Seyr, N., … Holzleitner, J. (2008), Computational analysis of injection- molding residual-stress development in direct-adhesion polymer-to- metal hybrid body-in-white components. *Journal of Materials Processing Technology*, 203(1–3), 19–36. doi:10.1016/j.jmatprotec.2007.09.059.

14. Stoić, A., Kopac, J., Duspara, M., Micetic, I., & Stoic, M. (2013), Manufacturing of injection moulding tool with five axis milling machine. *Journal of Achievements in Materials and Manufacturing Engineering*, 58(01), 38–46.

15. Smith, A. G., Wrobel, L. C., McCalla, B. A., Allan, P. S., & Hornsby, P. R. (2008), A computational model for the cooling phase of injection moulding. *Journal of Materials Processing Technology*, 195(1–3), 305–313. doi:10.1016/j.jmatprotec.2007.05.018.

16. Chen, C.-P., Chuang, M.-T., Hsiao, Y.-H., Yang, Y.-K., & Tsai, C.-H. (2009), Simulation and experimental study in determining injection molding process parameters for thin-shell plastic parts via design of experiments analysis. *Expert Systems with Applications*, 36(7), 10752–10759. doi:10.1016/j.eswa.2009.02.017.

17. Product Information, Ultramid® A3WG6 BK00564 Polyamide 66 (2015), BASF Corporation, Engineering Plastics,1609 Biddle Avenue,Wyandotte, MI 48192.

18. Taha, I., & Abdin, Y. F. (2011), Modeling of strength and stiffness of short randomly oriented glass fiber—polypropylene composites. *Journal of Composite Materials*, 45(17), 1805–1821. doi:10.1177/0021998310389089.

19. El Otmani, R., Zinet, M., Boutaous, M., & Benhadid, H. (2011), Numerical simulation and thermal analysis of the filling stage in the injection molding process: Role of the mold-polymer interface. *Journal of Applied Polymer Science*, 121(3), 1579–1592. doi:10.1002/app.33699.

20. Barriere, T., Liu, B., & Gelin, J. C. (2003), Determination of the optimal process parameters in metal injection molding from experiments and numerical modeling. *Journal of Materials Processing Technology*, 143–144, 636–644. doi:10.1016/s0924-0136(03)00473-4.

21. Ozcelik, B., & Sonat, I. (2009) Warpage and structural analysis of thin shell plastic in the plastic injection molding. *Materials & Design*, 30(2), 367–375. doi:10.1016/j.matdes.2008.04.053.

22. Ghose, A., Montero, M., & Odell, D., (2015), Characterization of an injection molding process for improved part quality, Berkeley Manufacturing Institute, Dept. of Mechanical Engineering, University of California - Berkeley, 3 (2), 31–38.

23. Sykutera, D., & Bieliński, M. (2012), Application of CA systems at Design and simulation of plastic molded parts, *Journal of Polish CIMAC*, 4(1), 65–72.

24. Chun-Ying, Z., & Li-Tao, W., (2011), Injection mold design based on plastic advisor analysis software in Pro/E, *International Conference*, China, IEEE, 205–208, doi:10.1109/CMSP.2011.49.

25. Zhou, H., Yan, B., & Zhang, Y. (2008), 3D filling simulation of injection molding based on the PG method. *Journal of Materials Processing Technology*, 204(1–3), 475–480. doi:10.1016/j.jmatprotec.2008.03.017.

26. Kirchberg, S., Holländer, U., Möhwald, K., Ziegmann, G., & Bach, F.-W. (2012), Processing and characterization of injection moldable polymer-particle composites applicable in brazing processes. *Journal of Applied Polymer Science*, 129(4), 1669–1677. doi:10.1002/app.38862.

27. Panda, S. B., Nayak, N. C., & Mishra, A., (2017), Engineering polymers in automobile seat belt lock applications: It's development, investigation and performance analysis. *Journal of Production Engineering*, 20(1), 63–68. doi:10.24867/JPE-2017-01-063.

28. Kovacs, J. G., & Siklo, B. (2011), Investigation of cooling effect at corners in injection molding. *International Communications in Heat and Mass Transfer*, 38(10), 1330–1334. doi:10.1016/j.icheatmasstransfer.2011.08.007.

29. Frick, A., & Spadaro, M. (2017), Mold design for the assembly injection molding of a solid housing with integrated dynamic seal. *Polymer Engineering & Science*, 58(4), 545–551. doi:10.1002/pen.24766.

30. Villarreal, M. G., Castro, M. J. M., & Cabrera-Ríos, M. (2011), A multicriteria simulation optimization method for injection molding, *Proceedings of the 2011 Winter Simulation Conference*, Arizona, USA, 2390–2402.

5 Effect of Adaptive Depth-First Sphere Decoding Scheme to MIMO-OFDM System in FSO

Chinmayee Panda and Urmila Bhanja
Indira Gandhi Institute of Technology

CONTENTS

5.1 INTRODUCTION

Free space optical (FSO) communication plays a vital role in the current research area. It is a technology where light travels in free space from transmitter to receiver end. Various turbulent factors such as rain, fog, and associated noises are the obstacles in transmitting the signal. Again, beam scattering and scintillation are the other causes of not getting the perfect signal at the receiver end. Hence, the receiver design is very important concern through which the signal degradation can be minimized [1]. Different modulation techniques such as binary phase-shift keying, quadrature amplitude modulation (QAM), and quadrature phase-shift keying are applied for long-distance communication [2]. The recent techniques like orthogonal frequency division multiplexing (OFDM) and multiple input multiple output (MIMO) are associated with FSO to increase the bit error rate (BER) performance and to reduce the receiver complexity [3]. OFDM is the technique where the orthogonal property of subcarriers helps to save the bandwidth, and ISI cancellation is the another advantage of OFDM. In MIMO technique, a more number of input antennas and a more number of output antennas are used for avoiding multipath fading. By applying MIMO technique, the spectral efficiency and system capacity increase through which a strengthened output is obtained [4]. When the combined techniques of MIMO and OFDM are applied to free space, the BER performance increased and the receiver complexity reduced with less power consumption. Green technology refers to the practical application of technology which is friendly to the environment, functioning in non-polluting ways, or regarded as less harmful to the environment. Various factors such as range of activities, methods, materials, and techniques like energy recycling, renewable energy sources, and energy efficiency are included in green technology [5]. In this chapter, a new technology is discussed that reduces BER at the receiver end, diminishes the system complexity, and increases the power efficiency, which are the factors of green technology.

The detection schemes associated with MIMO-OFDM are zero-forcing (ZF) and maximum-likelihood (ML) detection schemes. ML detection is more complex when the number of transmitting antennas increases. Hence, depth-first sphere decoding (DFSD), which uses non-linear detection algorithm, is proposed. The DFSD detection scheme has better performance and is lesser complex as compared with the ML detection scheme [6]. But the DFSD detection scheme shows high complexity in its hardware implementation. This disadvantage can be overcome by an adaptive DFSD (ADFSD) detection scheme, which shows low complexity than the classical DFSD detection scheme. The ADFSD detection scheme utilizes thresholds based on channel condition to deviate the number of top layer nodes in DFSD tree structure [7].

The chapter is organized as follows. Section 5.1 gives the introduction, Section 5.2 illustrates system model, Section 5.3 classifies the different detection schemes with the applied channel model, Section 5.4 shows the simulation result, and Section 5.5 summarizes the chapter.

5.2 SYSTEM MODEL

The channel model for MIMO-OFDM is shown in Figure 5.1. Here blocks of data are sent to OFDM modulator where cyclic prefix insertion and inverse fast Fourier transform process followed by serial to parallel conversion take place and the blocks of data are given to MIMO encoder. After the encoding process, the data are passed through the free space and then decoded at the receiver end by MIMO decoder. The decoded data are passed through OFDM demodulator where the reverse process of modulation takes place and original data with associated noise are obtained at the receiver side.

The equivalent model for MIMO-OFDM is given as $R = HS + N$, where $R = [r1, r2,..., rNr]$ and T is the $Nr \times 1$ receive symbol vector; $S = [s1, s2,..., sNt]$ and T is the $Nt \times 1$ transmit symbol vector; and N is the $Nr \times 1$ noise vector. $[\cdot]T$ indicates the transpose matrix. Equation (5.1) represents matrix H [8]

$$H = \begin{pmatrix} h11 & \cdots & h1Nt \\ \vdots & \ddots & \vdots \\ hNr1 & \cdots & hNrNt \end{pmatrix} \qquad (5.1)$$

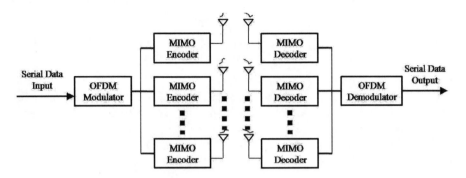

FIGURE 5.1   MIMO-OFDM-FSO system.

where h_{ij} ($i = 1, 2,\ldots, Nr$), ($j = 1, 2,\ldots, Nt$) indicates channel coefficient from the jth transmitting antenna to the ith receiving antenna.

5.3   VARIOUS DETECTION SCHEMES AND CHANNEL MODELLING

5.3.1   MAXIMUM LIKELIHOOD DETECTION

The general equation for ML detection scheme is defined by Equation (5.2) [9]

$$S_{ML} = \arg_t \min\left\{\|R - Hs\|^2\right\} \text{ and } t = x\varepsilon |S|^{Nt} \tag{5.2}$$

where S denotes the set of constellation points of transmitter symbol and $|S|^{Nt}$ denotes space of transmitter vector. This detection scheme has optimum error performance, and hence, it is used at receiver side, but the complexity increases by increasing the number of transmitting antennas.

5.3.2   CLASSICAL DEPTH-FIRST SPHERE DECODING

Classical DFSD is like ML detection scheme having low complexity. The DFSD detection scheme inequality is given by Equation (5.3) [10]

$$\|R - Hs\|^2 \le C^2 \tag{5.3}$$

where C is the sphere radius. By representing this decoding scheme in a tree structure, we can analyze it very easily. The equation of channel matrix is given in Equation (5.4)

$$H = IT \tag{5.4}$$

where I is the $Nr \times Nr$ unitary matrix, T is the $Nr \times Nt$ upper triangular matrix, $Z = I^H R = I^H (HS+N) = TS + I^H N$, and in matrix form Z is given by Equation (5.5) [11]

$$Z = \begin{pmatrix} r11 & \cdots & r1Nt \\ \vdots & \ddots & \vdots \\ 0 & \cdots & 0 \end{pmatrix} S + N \tag{5.5}$$

where $(.)^H$ indicates the Hermitian transpose, Z indicates the received signal vector, and N denotes the noise vector. The main purpose of DFSD detection scheme includes determination of the first radius in depth direction from the tree structure. A squared Euclidean distance (ED) is used in this detection scheme.

5.3.3 ADAPTIVE DEPTH-FIRST SPHERE DECODING

ADFSD has low complexity than the ML detection scheme. The ADFSD detection scheme is evaluated depending upon the number of nodes at the Nt-th layer. By considering the overall adaptive DFSD tree structure with MIMO-OFDM system having QAM modulation and by taking the ascending order of squared ED, the nodes are sorted at the Nt-th layer in the ADFSD detection scheme depending on the turbulence conditions of free space.

In ADFSD scheme, the first radius is calculated in Equation (5.6) [7]

$$D_1^1(1) = \left\| z_1 - \left\{ \mathrm{rad}_{11} U(1) + \mathrm{rad}_{12\,x_2^1} + \mathrm{rad}_{13\,x_3^3} + \mathrm{rad}_{14\,x_4^1} \right\} \right\|^2 + D_2^3(1) \tag{5.6}$$

The reference symbol is indicated by $U(k)$ $(k = 1, 2\ldots, u)$ and x is the t-th estimated symbol among the reference symbols from the Nt-th layer to the $(n + 1)$-th layer.

In ADFSD scheme, the shortest accumulated squared ED is given in Equation (5.7)

$$D_1^4(3) = \left\| z1 - \left\{ \mathrm{rad}_{11} U(3) + \mathrm{rad}_{12\,x_2^4} + \mathrm{rad}_{13\,x_3^4} + \mathrm{rad}_{14\,x_4^4} \right\} \right\|^2 + D_2^1(4) \tag{5.7}$$

The ADFSD detection scheme, utilizing all nodes at the Nt-th layer, predicts the signal more correctly than the DFSD detection scheme using one node at the Nt-th layer. In the appropriate channel condition, this detection scheme uses single node at the Nt-th layer. But, in turbulent conditions the ADFSD scheme uses entire nodes at the Nt-th layer.

5.3.4 CHANNEL MODELLING

The probability density function of log-normal distribution function is defined in Equation (5.8) [12]

$$f(x; \zeta, \xi) = \frac{1}{x \xi \sqrt{2\pi}} \cdot e^{\frac{-(\ln x - \zeta)^2}{2\xi^2}} \tag{5.8}$$

where the location parameter and the shape parameters are ζ and ξ, respectively.

5.4 SIMULATION RESULT

Table 5.1 lists the parameters taken for simulation. Here 512 number of subcarriers are taken. The cyclic prefix is taken 32 with FFT size 128. The modulation used is 16-QAM with 2×2 and 4×4 MIMO systems. Different turbulence conditions such as strong and weak turbulence conditions are taken to simulate the results.

Figure 5.2 shows signal-to-noise ratio (SNR) versus BER plot of MIMO-OFDM system with DFSD and ADFSD schemes in FSO. It is verified that classical DFSD scheme shows a BER of 10^{-4} and ADFSD scheme shows a BER of 10^{-4} in wireless medium. In contrast, ADFSD scheme with MIMO-OFDM in FSO exhibits a BER of 10^{-5} and 10^{-6} in strong and weak turbulence conditions, respectively.

Figure 5.3 shows how the number of complex multiplications varies with the number of transmitting antennas. It is seen that with an increase in the number of

TABLE 5.1
Simulation Parameters [7]

Serial Number	Parameter	Value
1.	Number of subcarriers	512
2.	Modulation scheme	16-QAM
3.	Cyclic prefix	32
4.	FFT size	128
5.	Antenna configurations	2×2 and 4×4 MIMO
6.	C_n^2 (Weak turbulence)	$15 \times 10^{-15} \mathrm{m}^{-2/3}$
7.	C_n^2 (Strong turbulence)	$15 \times 10^{-13} \mathrm{m}^{-2/3}$

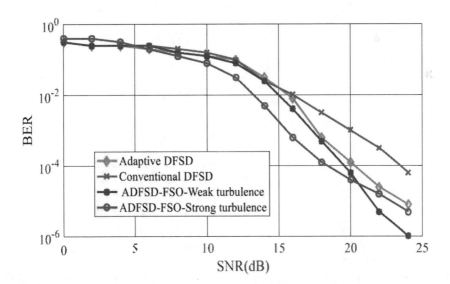

FIGURE 5.2 SNR versus BER plot of MIMO-OFDM system with DFSD and ADFSD schemes in FSO.

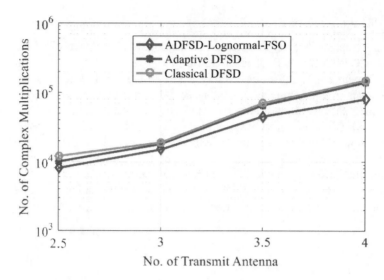

FIGURE 5.3 Number of transmit antennas versus number of complex multiplications for MIMO-OFDM.

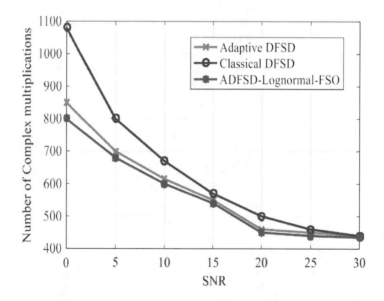

FIGURE 5.4 Comparative analysis of MIMO-OFDM system with DFSD and ADFSD.

transmitting antennas, the number of complex multiplications increases (1.5×10^5) for ADFSD-MIMO-OFDM-FSO scheme as compared with classical DFSD-MIMO-OFDM and ADFSD-MIMO-OFDM in wireless medium.

Figure 5.4 shows the analysis of 2×2 MIMO-OFDM-FSO system depending upon complexity and SNR. The FSO ADFSD scheme with MIMO-OFDM in

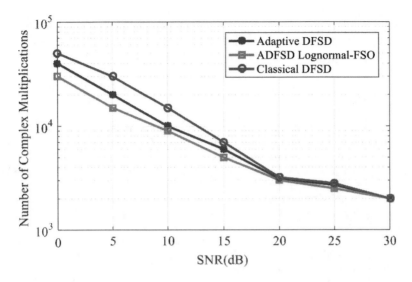

FIGURE 5.5 Comparative analysis of MIMO-OFDM system with DFSD and ADFSD.

log-normal channel shows lesser complexity compared with classical DFSD and ADFSD in wireless medium. MIMO-OFDM-FSO system with ADFSD scheme shows 440 number of complex multiplications for 20 dB SNR, whereas classical DFSD scheme shows 460, and ADFSD in wireless medium shows 450 number of complex multiplications.

Figure 5.5 shows comparative analysis of 4×4 MIMO-OFDM system with DFSD and ADFSD schemes. It is observed that ADFSD scheme and classical DFSD scheme show 5×10^4 and 4×10^4 numbers of multiplications in wireless medium, respectively. In contrast, by using log-normal channel, ADFSD scheme in FSO shows a BER of 3×10^4 number of multiplications. Hence, the complexity of the MIMO-OFDM system is reduced by using the ADFSD scheme in FSO.

5.5 CONCLUSION

This chapter uses the ADFSD detection scheme with MIMO-OFDM system in FSO to minimize BER and complexity of the system. This detection scheme is approximately equivalent to the ML detection scheme and the classical DFSD detection scheme. By taking log-normal channel in FSO, this scheme is compared with classical DFSD and ADFSD schemes. Again, by applying this detection scheme to FSO, complexity is minimized based on the number of transmitting antennas compared with the classical DFSD detection scheme. In strong and weak turbulence conditions, the ADFSD scheme exhibits a BER of 10^{-5} and 10^{-6}, respectively. This detection scheme is a useful detection scheme in FSO to minimize the power utilization and complexity which is a part of green technology.

REFERENCES

1. A. Kora, R. Hontinfinde and T. Ouattara, Free space optics attenuation model for visibilities ranging from 9 to 12 km, *Procedia Computer Science*, vol. 56, pp. 260–265, 2015.
2. G. Kaur, H. Singh and A. S. Sappal, Free space optical using different modulation techniques: A review, *International Journal of Engineering Trends and Technology (IJETT)*, vol. 43, no. 2, pp. 109–115, 2017.
3. J. Zhang, A. Sayeed and B. Van Veen, Low complexity MIMO receiver with decoupled detection, *Sensor Array and Multichannel Signal Processing Workshop Proceedings*, University of Wisconsin–Madison, 2002.
4. M. Jiang and L. Hanzo, Multiuser MIMO-OFDM for next-generation wireless systems, *Proceedings of the IEEE*, vol. 95, no. 7, pp. 1430–1469, 2007.
5. W. Chu, D. Chun and S. Ahn, Research advancement of green technologies, *International Journal of Precision Engineering and Manufacturing*, vol. 15, no. 6, pp. 73–977, 2014.
6. W. You, L. Yi and W. Hu, Reduced complexity maximum-likelihood detection for MIMO-OFDM systems, *8th International Conference on Wireless Communications, Networking and Mobile Computing*, USA, 2013.
7. S. J. Shim, S. J. Choi and H. K. Song, Low power consumption signal detector based on adaptive DFSD in MIMO-OFDM systems, *Energies*, vol. 12, p. 599, 2019.
8. J. Gao, O. C. Ozdural, S. H. Ardalan and H. Liu, Performance modeling of MIMO OFDM systems via channel analysis, *IEEE Transactions on Wireless communication*, vol. 5, no. 9, pp. 2358–2362, 2006.
9. J. Yue, K. J. Kim, J. Gibson and R. A. Iltis, Channel estimation and data detection for MIMO -OFDM systems, *IEEE Global Telecommunications Conference*, Arlington, VA, vol. 2, 2003.
10. Y. Nasser, S. Aubert, F. Nouvel, K. Y. Kabalan and H. A. Artail, A simplified hard output sphere decoder for large MIMO systems with the use of efficient search center and reduced domain neighborhood study, *EURASIP Journal on Wireless Communications and Networking*, vol. 2015, p. 227, 2015.
11. P. Athira and A. A. John, Channel matrix shaping scheme for MIMO OFDM system in wireless channel, *International Journal of Scientific & Engineering Research*, vol. 7, no. 7, pp. 139–146, 2016.
12. https://www.itl.nist.gov/div898/handbook/eda/section3/eda3669.htm.

6 Renewable Energy Adaptation Model for Sustainable Small Islands

Sanju Thomas
Cochin University of Science and Technology

Sudhansu S. Sahoo
College of Engineering and Technology

G. Ajith Kumar
Cochin University of Science and Technology

CONTENTS

6.1   INTRODUCTION

Small islands are among the most vulnerable regions in the world, since minimal changes in the climate and subsequent rising sea levels take an effect which threatens the livelihoods of entire regions with contamination of potable and agricultural water [1,2]. Climate change vulnerabilities and limited potential for economic growth are barriers for penetration of new technologies in small islands. Low population, volume of consumption of goods, and reach of supply chain limit new investments in small islands [3]. The island states are energy intensive, since the cumulative energy consumed by the island state/nation cannot be averaged across the total population [4]. Any change in oil prices will have an impact on the consumers, since its availability and sudden volatility can be felt through various verticals which include transport services, cost of energy, and cost of items transported from the mainland, including food and beverages [5,6]. It is seen that there will be a difference of 60%–100% unit cost between the main island and the other islands in case of archipelagos [7].

Considerable investments have been made in renewable energy (RE) sector in the small islands in form of technologies to tackle the volatility of the oil prices, focused on to replace oil-based power generation [4]. In spite of funding, the initiatives taken in small islands to promote RE often fail due to either technology or the method of implementation or follow up, including maintenance. Studies on evaluation of effectiveness of RE projects implemented on islands of Pacific Ocean reveal a stretch of failures because of indigenous technically qualified people available every time a maintenance issue was raised, which shows lack of frameworks [8,9]. However, recently technology and implementation strategies have improved the effectiveness of such projects [10]. Investigation on earlier failures reveal lack of policy frameworks, improper funding structure, and lack of indigenously available trained manpower are the main reasons for failure of such initiatives [11,12].

While developing financial models for RE in small islands, it is ideal to identify solutions with multiple benefits [13]. RE integration at large scale has become technically optimized with rooftop and ground mounted photovoltaic (PV) systems. However, investments on mini/microscale integrations to livelihood activities are necessary for small islands, which will reduce generation cost and avoid loss in returns [14]. Studies predict that a mix of different RE technologies will be more economical and suitable, and future projects should carefully consider the local conditions and support the technology most appropriate to the specific community. Another opportunity for RE integration in small islands is substituting the fuel for transportation sector, either through solar powered batteries or bio-fuels from locally available agricultural products [15]. Earlier studies have established that while mapping the potential of RE integration for small islands, integration of technologies that can use available local resources should be the primary focus [16]. Research work has also been done on the triangular approach, which considers the three interrelated dimensions of sustainability, namely, economic, social, and environmental [17]. RE integration and transition from conventional sources of energy into renewable should take into consideration the islands' socioeconomic aspects and sustainability strategies. Previous studies on RE integration in small islands have not considered the following major areas in analysis [17]:

- The process to make sure that RE resource will have reliability as primary source;
- An effective analysis on the payback procedure on the investments made in RE projects;
- Water usage, contamination of soil, and effects on land occupancy;
- Renewable energy sources (RES) promotion policies in islands;
- Contribution of RES to desalination;
- Possibilities of integrating with multiple RES or with existing conventional energy resources (hybridization);
- Designing the RES with an option to act as the primary source and meet the peak load;
- Research and analysis for application of integrating RES on a case-to-case basis.

Literature study shows that there are no specific guidelines established for adaptation of renewable sources as a primary/secondary energy source in small islands. Research also points out that small islands are vulnerable to the very slightest change related to global warming. The current study focuses on five principles that must be considered while considering adaptation of RE in small islands. A renewable energy adaptation model (REAM) is created for analyzing the effect of RE integration for the benefit of sustainable small islands. The REAM works on five principles to meet the sustainable development goals of small islands. The following methodology is followed: (1) development of REAM based on current limitations, (2) selection of a small island to implement the model for analysis, (3) defining two innovative projects for RE integration, (4) implementation of the REAM model on the selected projects, and (5) selection of a better alternative from the available options.

6.2 RENEWABLE ENERGY ADAPTATION MODEL (REAM) FOR SMALL ISLANDS

In accordance with our understanding of available literature and limitations of research on specific RE integration strategies for small island nations, the REAM is developed based on five sustainable principles, which are (1) reliability of the new technology,

(2) independency of the new technology, (3) contribution to the power–water–food nexus, (4) hybridization possibilities with existing resources, and (5) conservation of the ecology and environment. The objective of REAM is to analyze the sustainability of a project or technology and its adaptability with the sustainability principles of the small islands, through the lifecycle of the project. The REAM also reviews the five economies of scale of the new technology, which are (1) scalability to capacities, (2) commercialization, (3) generation of revenue to recover Capex, (4) vulnerability to externalities, and (5) livelihood improvements of the local population. REAM understands these economies of scale as very relevant in selection of technology and adaptation to sustainability principles of small islands. Earlier research proves that the lack of proper study on these economies of scale have led to lack of commercial returns leading to hindrances for successful business models [9,10].

6.3 REAM PRINCIPLES

6.3.1 RELIABILITY OF NEW TECHNOLOGY

The islands are usually remotely located from the mainland and require more than few hours to reach for aid in emergency. Hence, the reliability of the RE systems should be very high, with immediate alternatives to meet emergency. Usually diesel generators are used as primary source of energy with substantial backup capacities to address power failure. RE resources vary during time and season. The possibilities of hybrid energy integration with different/multiple sources of renewable and non-RE sources should be given importance than depending on a single RE source. Systems that have been tested and proven with satisfactory operations in the mainland should be deployed with special care given for the island conditions.

6.3.2 INDEPENDENCY OF THE NEW TECHNOLOGY

The new technology should be very independent from the mainland in its daily operational functions, routine maintenance, and decision-making processes. The power distribution mechanism should be independent from the mainland system with respect to infrastructure sharing or control architecture. Remote monitoring of the system should help replenish the spares for anticipatory breakdowns and predictive maintenance. Enough spares should be made available in the island with trained manpower capable of making the downtime of the system very little. The system should be integrated with independent dismountable components for easy dismantling and shipping to the mainland in case of a very serious damage.

6.3.3 CONTRIBUTION TO POWER–WATER–FOOD NEXUS

The power–water–food requirement and dependency are interrelated and complex for small islands, considering the area and effect of externalities of one project over the other. Hence, the effect on any one has an immediate unprecedented effect on the other.

The main objective of RE integrated projects in islands should be to reach the masses, in multiple spectrums of their daily life, without disturbing the existing sustainability balance. The RE integration should benefit multiple sectors contributing to the environmental sustainability and socioeconomic development of island community.

6.3.4   HYBRIDIZATION POSSIBILITIES WITH EXISTING RESOURCES

Any RE integration should be with an objective for possible hybridization with existing resources catering the power–water–food nexus. This can be in the form of load sharing during the peak hours, off-peak hours, and seasonal variations. The design of the system should take into consideration all possible hybrid combinations to meet even the partial load without disturbing the ecological balance. Even hybrid systems with integration of conventional energy sources should be considered.

6.3.5   CONSERVATION OF ECOLOGY AND ENVIRONMENT

The islands are the front-runners to face the impact of global climate change. Many of the small islands are situated on coral reefs and are vulnerable to slight deformations and imbalances. The local agriculture and livelihood mechanisms play a major role in the existence of the islands. The coral reef and the earth bed play a major role in the flow of underground water, thereby controlling the salinity of the potable water. Any new RE-based design should give utmost priority to safeguard the ecological and environmental balance.

6.3.6   LAKSHADWEEP: AN ARCHIPELAGO OF SMALL ISLANDS

Lakshadweep archipelago lies between latitude 8°30′ north to 12°30′ north and longitude of 71° east and 74° east in the Arabian Sea as a group of 36 islands out of which 11 are inhabited. The distance from Kochi, the nearest airport in the mainland, is approximately 400 km [18].

6.3.7   WATER SCENARIO

Ground water is replenished during the rainy season and is the only source for potable water. There is always a mixing between the sweet water and the saline water due to faults in pumping system or due to non-availability of enough head of sweet water. Since there is no proper drainage facility, the ground water gets contaminated. Earlier studies on the island have concluded that no single system or approach is capable to provide uninterrupted water, due to the typical geological and hydrogeological nature of these islands. A few islands have desalination plants running on diesel generators. Additional desalination plants and rainwater harvesting are the only viable options [19].

The ground water is very salty because of the heavy and frequent pumping of the ground water and also because of the sea water entering the ground water and contaminating the wells [20].

6.3.8 POWER SCENARIO

The diesel-run generators are the only power source in the island, without any interconnection grids between the islands. Diesel must be transported from the mainland, with spillages during the transportation being the main constraint. The power generation cost is very high with heavy subsidies given to the users. The distribution is through 11 kV and 415 V systems. The per square kilometer consumption of electricity in the island is higher than the mainland as the population density is high. Transmission losses are less because of less distance from generation to the final utility levels. The demand of electricity is mainly for the domestic applications, while industrial and agricultural sector loads are very less. Since the diesel is coming from mainland, the price remains high and the mode of transportation is hectic. To increase the reliability, diesel is stored well in advance. However, the soil being very porous transmits the spillage to the underground water bodies and contaminates them, which not only ends up in wastage but also affects the ecology of the land [20].

6.3.9 FOOD SCENARIO

The villagers depend on rice, cereals, and vegetables transported from the mainland for food. Coconut and byproducts are an active ingredient in the food staple. The egg and meat from the poultry farm are another resource of food. Continued efforts by the government to encourage dairy farming have made milk and milk byproducts largely self-sufficient. The villagers depend on verities of fish as an integral part of their diet. During the monsoon, they depend on the fish from the coral reef, while during the off-monsoon period, fishing is done from deep sea [20].

6.3.10 POWER–WATER–FOOD NEXUS

The detailing on the power, water, and food staple shows that they are interrelated. Desalination plants operate on diesel-run generators, which are also the source of power. Deep wells are not possible due to salinity. Fishing is the mainstay of economy, while mainland provides cereals and vegetables. Due to the land dimensions and high population index, the effect on any component of the power–water–food nexus will have an impact on the other. The RE integration projects should be designed to have a positive impact on all the three components of the nexus while carefully curtailing any externality effect on existing sustainability balance. Careful study of socioeconomic, environmental, and ecological aspects of the affected domains is required before the final design is selected.

6.4 AVAILABILITY OF RENEWABLE ENERGY RESOURCES AND ECOLOGY

6.4.1 SOLAR ENERGY

The annual average direct normal irradiance in the island is 4.032 kWh/m^2 · day [21]. The approximate land area required for ground-mounted solar PV installation is 5 acres per megawatt of power, while rooftops require comparatively more area. The major limitation for implementing solar PV is the unavailability of suitable land by adminis-tration. The land is largely held by local population, who prefer coconut farming and other livelihood mechanisms. The other limitation with the solar PV is the shade due to coconut leaves. The coconut trees grow tall, while most of the buildings are single-story buildings with tile roofs [21]. Power generation by solar thermal energy requires large land masses for concentrating mirrors, heavy structures for support and associated constraints logistics, and loading and unloading at coral reef shores of the island [18].

6.4.2   Wind Energy

Earlier studies on the wind potential for the island shows that the average wind speed during the monsoon (June, July, August) is 5.5–10.2 m/s. The pre-monsoon and post-monsoon wind speeds are estimated as 3–4.2 m/s. The analysis was conducted for the island of Kavaratti. The wind potential studies show that enough power for the island can be generated by the available wind potential of the island. The main limitations of the wind energy in the island are the ability to unload heavy equipment on vulnerable coral reef shores, with restrictions limited to 2 tons. This limits the machines to a capac-ity below 80 kW. The other major constraint is the mast height, which should be above the height of the coconut trees, with limits on the depth of the foundations [18,21].

6.4.3   Biomass Energy

Coconut is the major agricultural product of the islands. The husk, shell, and the cadjins of the tree are rich sources of biomass energy for power generation using gasification process. The approximate yield of coconuts per tree is estimated as 40 per year and the number of cadjins is 12. The biomass quantities are 0.32 kg/husk, 0.075 kg/shell, and 2 kg/cadjin. The approximate power generation potential from the palm trees in the island per year is estimated as 80 million kWh/year [18,20]. The major constraints in biomass gasification process are sizing of capacity, reliability of maintaining the required temperature, and pressure and exhaust. Even though, the biomass gasification process cannot replace the diesel generators altogether, but a substitution of up to 70% for diesel generators is reached [20].

6.4.4   Ecology and Environment

The island is stationed on coral reef with the ecosystem being a hub for a variety of fishes, especially ornamental fish of multiple species. The island depends on extracts from palm trees (coconut trees) as main source of agriculture output which are considered a commercial product. Population density, tourism, and lack of aware-ness have made the ecosystem vulnerable with accumulation of oil slicks, untreated domestic sewage, and solid waste [20,21].

6.4.5   Economy

The island is made to rely on available land characteristics (soil) and the surround-ing water to meet the economic need. The economy is dependent on locally grown agriculture products, animal husbandry and fisheries, the commercially available products being tuna, coir, vinegar, and copra [20].

6.4.6   MAJOR CONSTRAINTS OF THE ISLAND FOR SUSTAINABLE POWER–WATER–FOOD NEXUS

The following are the major constraints of Lakshadweep Islands [19]:

- Intrusion of saline water into domestic wells due to over extraction of potable water.
- Dependency on mainland for all requirements, which reflects on fluctuation in commercial/daily-use products.
- The islands are dependent on conventional power resources, especially the petroleum byproducts for industry and transport sector.
- The land pattern and its availability do not allow large-scale integration of RES.
- Resource harvest from the reefs has brought many reefs to various degrees of stress.
  - There is no ground water resource, and potability is dependent on rainfall.
  - Dependence on isolated desalination plants for supply of safe drinking water.
  - Diesel as the main source of electricity generation.
  - Lack of options in multiple livelihood activities.

6.4.7   MOST PRACTICED RENEWABLE ENERGY INTEGRATIONS

Considering the area constraint of small islands, RE integrations in small islands should be entirely different from mainland. Solar energy is considered as clean energy and do not require stringent environment clearances. Environmental impact studies suggest impact mitigation measures which are easy to implement in mainland. However, small islands will have constraints, considering the externalities of these mitigation measures. The most common RE integrations and their corresponding constraints for implementation in Lakshadweep archipelago are mentioned in Table 6.1.

6.4.8   INNOVATIVE PROJECTS FOR RENEWABLE ENERGY INTEGRATION

Taking the above constraints of the island and limitations of integrating conventional methods of RES, two innovative projects are discussed:

1. Seagoing vessels are mounted with solar PV panels and battery energy storage systems (BESS). The vessels offload the power to battery systems/diesel-run grids at wharfs.
2. Standalone microgrids, with combination of solar PV, small wind, biomass gasifier, and BESS, operating as primary source of power.

TABLE 6.1

Limitation for Common Renewable Energy Implementation Technologies in Lakshadweep

Common Renewable Energy Integration Methods	Constraints in Lakshadweep Archipelago
Grid-integrated ground-mounted solar PV	Availability of suitable land with clear titles and free of shadows
Grid-integrated rooftop-mounted solar PV	Availability of rooftop space, free of shadows. Most of the roofs are made up of thatched tiles. RCC roofs are given priority for storage of water
Solar thermal-based power/heat	Fabrication of heavy structures in the island, logistics of heavy structures to the island, availability of enough land free of shadows, wind loads
Large wind turbines	The mast of the turbines should be taller than the coconut trees. Foundation for the mast is a constraint with soil bearing capacity
Biomass gasifier	Although the availability of biomass from coconut trees is abundant, a proper supply chain for collection, storage, and usage must be established. Since the region has intense monsoon, biomass gasifier cannot be a single source of power

6.4.8.1 Solar Panels Mounted on Top of Seagoing Vessels with BESS

Lakshadweep Islands depend on fishing as the main stay of economy. Transportation of men and material between the islands and mainland is dependent on ships, ferries, and boats. Irrespective of being on trip or not, these modes of transport have to remain on water or afloat at wharfs/docks. There is enough space on top of these boats, ferries, and ships to hold a minimum number of panels. The panels can be customized based on the area available on decks/roofs of these water transport systems. The panels mounted on decks/roofs will generate power and store in BESS to offload them to main battery storage or diesel grids at the wharfs while the boats return. The local administration takes care of storage and distribution through existing grid network. The vessel owners are paid by the local administration based on the number of units of power offloaded to the BESS at the wharfs. Thus, a sustainable business model can be established.

In this project, RE is proposed as a secondary source of power to complement the primary source of energy, which is diesel generator–run grid.

6.4.8.2 Standalone Microgrids, with Combination of Solar PV, Small Wind, Biomass Gasifier, and BESS, Operating as Primary Source of Power

Land/rooftop-based grid-connected solar PV systems have limitations in islands due to availability of land and transmission networks. Standalone small units of RE- based power units sized for small domestic capacities will be a better option. Since a single RES has limitation of availability, a combination of RESs can be a better option with BESS. The BESS will keep charging and discharging, based on availability of the RES. The microgrid can be sized to cater the power requirement of a maximum of five domestic requirements.

Solar panels can be mounted on roofs, while small windmills can be installed on sunshades, ground, or rooftops. The biomass can be collected, stored, and gasified for heat/power applications.

The collection and gasification of coconut shredding can be an effective way for solid waste disposal which will enhance tourism options. The operation of microgrid can be a successful business model, with agreed feed-in-tariff between the operator and the five families, who share the space for solar, wind, and biomass gasifier.

In this project, the microgrid is considered as the primary source of power, while the existing diesel generator–run grid will be the secondary source of power.

6.5 IMPLEMENTATION OF RENEWABLE ENERGY ADAPTATION MODEL (REAM) FOR SUSTAINABILITY ANALYSIS OF INNOVATIVE PROJECTS

The above options of RE integrations may look very optimistic. But the successful implementation and adaptation to local sustainability balance is assessed through REAM. The results of analysis are shown in Table 6.2. The economic analysis of the two innovative projects is done through the five economies of scale, namely scalability, commerciality, breakeven, vulnerability, and livelihood improvement. The outcome of analysis is shown in Table 6.3.

6.6 RESULTS AND DISCUSSION

The REAM analysis shows that standalone microgrid, with a combination of solar PV, small wind, and biomass gasification, is an enterprising solution for dissemination of RE-based systems in small islands, especially since it decomposes the local biomass, which is otherwise a limitation for disposal. However, the biomass gasification process is complex and becomes critical for small capacities. The other drawback with biomass gasification–based microgrid is the management of exhaust from gasification process. While the reliability of microgrids can be very high, with multiple RESs integrated to BESS, the design is very complex. Ramp-up/ramp-down to meet the load curves and switching between different RE resources and BESS is also complex. Changes in rooftop design to mount the solar panels and small wind units including clearance of land area for gasifier and storage of biomass is essential. A workable business model with mutually agreeable feed-in-tariffs can make the business model profitable.

Seagoing vessel–mounted solar panel project is comparatively simple in terms of design, capital expenses, and operational expenses. A few successful pilot initiatives will provide lessons for scale-up and commercial operations. The project does not require any land acquirement and development activities. The capital expenses for vessel-mounted solar panels are less. Unlike the microgrid program, the vessel-mounted solar PV does not have many additional infrastructure requirements at land which can imbalance sustainable equilibrium. Vessel-mounted solar panels are less complex and require a shorter time for implementation.

TABLE 6.2

Analysis of Innovative Renewable Energy Projects Through REAM

REAM Principles	Option 1			Option 2		
	Result	Critical Criteria	Adaptation Parameter	Result	Critical Criteria	Adaptation Procedure
Reliability of the technology	High	Efficiency of batteries on board vessel and at the wharfs.	Efficient battery management system.	High	Small biomass gasification units are not commercialized.	Effective sizing of solar, wind, and biomass gasifier.
		The responsiveness of vessel operators to collect, store, and transfer the power to the battery storage system.	The feed in tariff (FiT) for the vessel operators should be enterprising.		Balance of ramp-up and ramp-down of multiple renewable energy units.	
Independency of the system	High	Battery storage system in the vessels should be of variable capacities.	Standardize the capacities based on the vessel application and sail hours.	Medium	Renewable energy sources generate power at different hours of the day.	Sizing of each renewable energy-based unit to suit the capacity.
		The battery management system should have easy mount–dismount mechanism.	Standardize the BESS and FiT through policy interventions.		Synchronizing multiple sources to suit load curves.	Number of backup systems to meet the reliability requirements.
Contribution to power–water–food nexus	High	No critical parameters (the load on the generators can be reduced).	Revision of policies to implement the same and benefit local population.	Medium	Land acquirement/clearances for microgrid.	Policy from the administration in favor of microgrid.
possibilities of hybridization;	High	The efficiency of BESS.	Standardize the batteries at wharfs and in vessels.	High	The complexity of the system design	Policy from administration to promote microgrids
conservation of ecology/environment	High	More number of vessels to dock at wharfs.	Careful design of wharfs for multiple docking of vessels during power transfer.	High	Emission of smoke from biomass gasifier.	Policy interventions to promote biomass gasification to eliminate solid waste.

TABLE 6.3

Economics Analysis of Projects

Project Economic Factors	Option 1	Option 2
Scalability	The opportunities for scaling up are high. Pilot scale project required before scaling up.	The scaling up is dependent on the design optimization and successful business models.
Commercial ability	The project is commercial. The vessel operators can charge the administration for power offload to battery energy storage system/grid.	Commercial if design can be optimized. Reliability and quality of power will be deciding factors for the customers.
Breakeven	The capex is less. Investment only on seaworthy solar PV panels. High opex on seaworthy painting for module mounting structures. Breakeven will be dependent on vessel type and purpose.	Capex for microgrid will be comparatively very less. Opex will be high if biomass gasifier is incorporated as a component. FiT to cover breakeven of capex.
Vulnerability to externalities	Independent of externalities and vulnerabilities.	Supply chain for biomass needs to be established.
Impact on livelihood	If implemented properly, the project can improve the livelihood activities through additional income and more reliable power supply.	Successful microgrids can improve the livelihood activities of the local community.

The analysis shows that though standalone microgrid is practically possible as the primary source of energy, the same is not recommended, for small islands, due to the complexity of the design. The standalone microgrid would have been an ideal project in mainland, since any change in combinations could have been experimented, modified, and established for approval. The REAM analysis concludes that though both projects may look feasible for RE integration, considering the land occupancy rate, ecological impacts, project complexity, and externality factors, vessel-mounted solar panels seem a better proposal for implementation than the standalone microgrids.

6.7   CONCLUSIONS

Many small islands are largely dependent on diesel generators as the main source of energy. RESs are yet to replace conventional sources of energy, primarily because of scattered population in the islands and less rate of returns on investments. However, while integrating RE in small islands, the adaptation of the technology for sustainability balance of the island is very important. The power–water–food nexus in small islands is interconnected, with any positive change in one component probably adversely impacting the other components of the nexus with irreversible effects. Many efforts to implement RE projects in small islands were not successful, because of selection of wrong technologies, implementation strategies, or lack of policy regulatory measures.

REAM is created with five basic principles which are reliability, independency, contribution, hybridization, and conservation. The model is ideal for analysis of success in adaption of RE integrated projects in small islands. Two innovative RE integrated projects are assessed through REAM. It is concluded that very complex designs not only will have reliability issues but externality impacts too. The analysis shows that projects that do not disturb the larger sustainable balance of the island should be selected for RE integration.

REFERENCES

1. Nurse, L. A., et al. (2014), Small islands. In: Climate Change 2014: Impacts, Adaptation, and Vulnerability. Part B: Regional Aspects. *Contribution of Working Group II to the Fifth Assessment Report of the Intergovernmental Panel on Climate Change*.
2. Barros, V. R., et al., Eds. (2014), Impacts, adaptation, and vulnerability: Part B: Regional aspects. *Contribution of Working Group to the Fifth Assessment Report of the Intergovernmental Panel on Climate Change*; Cambridge University Press: Cambridge; New York.
3. Cottrell, J., Fortier, F., and Schlegelmilch, K. (2015), Fossil fuel to renewable energy: Comparator study of subsidy reforms and energy transitions in African and Indian Ocean Island States, United Nations Office for Sustainable Development, Incheon, Republic of Korea, January 2015.
4. Dornan, M. (2015), Renewable energy development in small island developing states of the pacific. *Resources*, 4, 490–506; doi:10.3390/resources4030490.
5. Fiji Electricity Authority. Annual Report; FEA: Suva, Fiji (2010)
6. Reserve Bank of Fiji. Quarterly Review: June 2011; Reserve Bank of Fiji: Suva, Fiji (2011)
7. Wade, H., et al. (2005), Pacific Regional Energy Assessment 2004: An Assessment of the Key Energy Issues, Barriers to the Development of Renewable Energy to Mitigate Climate Change, and Capacity Development Needs to Removing the Barriers. Prepared for the Secretariat of the Pacific Regional Environment Programme: Pacific Islands Renewable Energy Project, Vol. 1–16.
8. Mala, K., Schläpfer, A., and Pryor, T. (2008), Solar photovoltaic (PV) on atolls: Sustainable development of rural and remote communities in Kiribati. *Renewable and Sustainable Energy Reviews*, 12, 1345–1363.
9. Dornan, M. (2014), Access to electricity in small island developing states of the pacific: Issues and challenges. *Renewable and Sustainable Energy Reviews*, 31, 726–735.
10. Dornan, M. (2011), Solar-based rural electrification policy design: The renewable energy service company (Resco) model in Fiji. *Renewable Energy*, 36, 797–803.
11. IRENA Report (2012), Electricity storage and renewables for island power - A guide for decision makers.
12. Niles, K. (2013), *Energy Aid in Caribbean and Pacific Small Island Developing States (SIDS)*. University of Otago: Dunedin.
13. Island Energy – Status and Perspectives Workshop Summary Report 5 and 6 October 2015. Institute of Applied Energy Tokyo, Japan.
14. Dornan, M., and Jotzo, F. (2015), Renewable technologies and risk mitigation in small island developing states: Fiji's electricity sector. *Renewable and Sustainable Energy Reviews*, 48, 35–48.
15. Klock, C. (2016), Fueling the Pacific: Aid for renewable energy across Pacific Island countries. *Renewable and Sustainable Energy Reviews*, 58, 311–318.
16. Duic, N., et al. (2008), Renew islands methodology for sustainable energy and resource planning for islands. *Renewable and Sustainable Energy Reviews*, 12, 1032–1062.

17. Jaramillo-Nieves, L., and del Río, P. (2010), Contribution of renewable energy sources to the sustainable development of islands: An overview of the literature and a research agenda. *Sustainability*, 2, 783–811. doi:10.3390/su2030783.

18. Lakshadweep Action Plan on Climate Change (LAPCC) (2012), Department of Environment and Forestry Union Territory of Lakshadweep, Supported by UNDP, 2012 Available at http://moef.gov.in/wp- content/uploads/2017/08/Lakshadweep.pdf (Accessed on 10/10/2019)

19. The Energy and Research Institute (TERI) Report (2001), Strategy plan for 100% RET utilization in Lakshadweep island, The Energy Research Institute, TERI Report No. 2001 RT 41, Available at http://www.globalislands.net/userfiles/_indi a_lakshadweep1. pdf (Accessed on 20/10/2019)

20. Ansari, S. M., Manaullah, and Jalil, F. M. (2016), Investigation of renewable energy potential in Union Territory of Lakshadweep islands, Innovative applications of computational intelligence on power, energy and controls with their impact on humanity, CIPECH 2016.

21. Jinoj, T. S. P., et al. Analysis of renewable energy potential sources for Kavaratti island, Lakshadweep, *Proceedings of International Conference on Sustainable Environment and Civil Engineering*' in *IOSR Journal of Engineering*, ISSN 2278-8719.

7 Numerical Simulation of Circular Two-Phase Jet Flow and Heat Transfer

S.S. Bishoyi
Christian College of Engineering and Technology

P.K. Tripathy
Utkala Gourav Madhusudan Institute of Technology

CONTENTS

7.1 INTRODUCTION

Two-phase jet flow finds its various applications from the grassroots of engineering applications such as separating dust particles from the raw materials to the high-end applications of biomedical engineering. The tremendous growth in the application of jet flow analysis has made it mathematically applicable to all the fields of engineering studies. Computational fluid dynamics is the most basic mathematical model used to understand the property and the interaction of various particles such as water droplets and dust particles. It is used to optimize the equipment size of a particular plant while analyzing the laminar flow or turbulent flow of fluid [1]. Bounded linearized equations can be framed by perturbing the velocity distribution function which was studied by Schlichting [6]. This is only applicable to laminar fluid. In the subsequent time, Bansal and Tak studied the temperature distribution for the laminar fluid under nonfractional heating conditions [2]. Interference of dilute suspension particles and the corresponding Brownian diffusion have been studied by Soo where it was found that the interaction of considered suspension particle is very less in case of laminar flow. Soo has also demonstrated in his paper that spreading of particle is due to fluid motion, and hence the particles are settled at the base of the fluid due to its density, however the momenta has been dissipated and disturbed. Submicron particles have shown their movement when Brownian motion for the distributed concentration is significant. Applied transverse magnetic field and their stress effect have been studied by Chamkha. Panda et al. investigated the effect of fractional volume and heat transfer on the flow of fluid [3]. Analysis of small solid particles and their interaction with a two-dimensional plane laminar jet have been studied by Ryhming [4]. He cancelled the drag force acting on the particle during the analysis of fluid momentum. In contradiction, Dutta et al. have investigated the performance of momentum equation under the drag and transverse force [9]. Particle volume fraction becomes significant only when sufficiently large density ratio is applied to jet analysis. Under this condition, the pressure distribution inside the laminar flow due to the suspended particle can be neglected for particle diameters up to few hundreds as investigated by Rudinger [5]. This investigation clearly states that the contribution of pressure due to suspended particle in a random motion can be neglected. From the above literature review, it can be understood that two properties such as mass fraction and volume fraction exist under equilibrium condition where one is constant and other is variable, respectively. In an experiment, Dutta and Das considered incompressible dusty fluid which is both temperature and velocity invariant for both symmetrical and unsymmetrical jet mixing. Here, it has been considered that the particle is different from its surrounding stream in terms of both temperature and velocity as discussed earlier.

7.2 MATHEMATICAL FORMULATION

Let us consider the case of a fluid flowing from a small circular orifice in a wall and mixing with the same surrounding fluid at rest, where the pressure is constant, the axis of z is along the axis of the jet and r denotes distance from the axis. Since the pressure is constant in the flow field and motion is steady, z = constant, the rate at which momentum flows across a section, must be a constant. Figure 7.1 shows the stream line pattern for circular laminar jet.

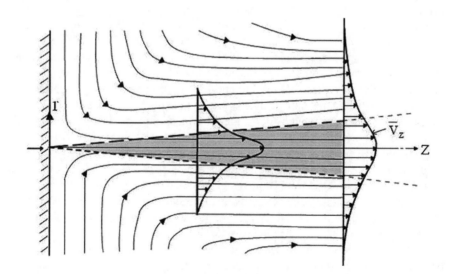

FIGURE 7.1   Stream line pattern for a circular laminar jet.

The boundary layer equations of the steady flow of incompressible fluid containing special purpose machine in cylindrical polar coordinates (r, θ, z) in axisymmetric case can be written as follows:

$$\frac{\partial}{\partial r}(rv) + \frac{\partial}{\partial z}(ru) = 0 \tag{7.1}$$

$$\frac{\partial}{\partial r}\left[\rho_p rv_p\right] + \frac{\partial}{\partial z}\left[\rho_p ru_p\right] = 0 \tag{7.2}$$

$$\frac{\partial p}{\partial r} \approx 0(\delta) \tag{7.3}$$

$$\left[v\frac{\partial u}{\partial r} + u\frac{\partial u}{\partial z}\right] = v\left(\frac{\partial^2 u}{\partial r^2} + \frac{1}{r}\frac{\partial u}{\partial r}\right) - \frac{\rho_p}{\rho}\frac{1}{\tau_p}\frac{1}{1-\varphi}(u - u_p) \tag{7.4}$$

$$\left[v_p\frac{\partial u_p}{\partial r} + u_p\frac{\partial u_p}{\partial z}\right] = v_p\left(\frac{\partial^2 u}{\partial r^2} + \frac{1}{r}\frac{\partial u}{\partial r}\right) + \frac{1}{\tau_p}(u - u_p) \tag{7.5}$$

$$\rho C_p\left[v\frac{\partial T}{\partial r} + u\frac{\partial T}{\partial z}\right] = \frac{\rho_s C_s \varphi(T_p - T)}{(1-\varphi)\tau_T} + K\left(\frac{\partial^2 T}{\partial r^2} + \frac{1}{r}\frac{\partial T}{\partial r}\right)$$
$$+ \mu\left(\frac{\partial u}{\partial r}\right)^2 + \frac{\rho_p}{\tau_p(1-\varphi)}(u - u_p) \tag{7.6}$$

$$\rho_p C_s\left[v_p\frac{\partial T_p}{\partial r} + u_p\frac{\partial T_p}{\partial z}\right] = K_p\left[\frac{\partial^2 T_p}{\partial r^2} + \frac{1}{r}\frac{\partial T_p}{\partial r}\right] - \frac{\rho_p C_s(T_p - T)}{\tau_T}$$
$$+ \frac{\partial}{\partial r}\left[u_p \varphi \mu_s \frac{\partial u_p}{\partial r}\right] - \frac{\rho_p}{\tau_p}\left[(u - u_p)^2\right] \tag{7.7}$$

The nondimensional forms of variables are given by

$$\bar{z} = \frac{z}{\lambda}, \bar{r} = \frac{r}{(\tau_p v)^{\frac{1}{2}}}, \bar{u} = \frac{u}{U}, \bar{v} = v\left(\frac{\tau_p}{v}\right)^{\frac{1}{2}}, \bar{u}_p = \frac{u_p}{U}, \bar{v}_p = v_p\left(\frac{\tau_p}{v}\right)^{\frac{1}{2}},$$

$$\bar{\rho}_p = \frac{\rho_p}{\rho_{p0}}, \bar{\mu}_p = \frac{\mu_p}{\mu_{p0}}, \bar{T} = \left(\frac{T - T_0}{T_0}\right), \bar{T}_p = \left(\frac{T_p - T_0}{T_0}\right), \bar{K}_s = \frac{K_s}{K_{s0}}, \tag{7.8}$$

$$Pr = \frac{\mu_0 C_{p0}}{K_o}, \lambda = \tau_p U, Ec = \frac{U^2}{C_p T_0}, \in = \frac{v_{p0}}{v_0}, D_p \approx v_p, \alpha = \frac{\varphi \rho_{s0}}{(1-\varphi)\rho}$$

Introduce nondimensional variables of Equation (7.8) in Equations (7.1)–(7.7) after dropping bars. The dimensionless equations are as follows:

$$\frac{\partial}{\partial r}(rv) + \frac{\partial}{\partial z}(ru) = 0 \tag{7.9}$$

$$\frac{\partial}{\partial r}\left[\rho_p r v_p\right] + \frac{\partial}{\partial z}\left[\rho_p r u_p\right] = 0 \tag{7.10}$$

$$\left[v\frac{\partial u}{\partial r} + u\frac{\partial u}{\partial z}\right] = \left(\frac{\partial^2 u}{\partial r^2} + \frac{1}{r}\frac{\partial u}{\partial r}\right) - \alpha\rho_p\left(u - u_p\right) \tag{7.11}$$

$$\left[v_p\frac{\partial u_p}{\partial r} + u_p\frac{\partial u_p}{\partial z}\right] = \epsilon\left(\frac{\partial^2 u_p}{\partial r^2} + \frac{1}{r}\frac{\partial u_p}{\partial r}\right) - \left(u - u_p\right) \tag{7.12}$$

$$\left[v\frac{\partial T}{\partial r} + u\frac{\partial T}{\partial z}\right] = \frac{2\alpha}{3Pr}\rho_p\left(T_p - T\right) + \frac{1}{Pr}\left(\frac{\partial^2 T}{\partial r^2} + \frac{1}{r}\frac{\partial T}{\partial r}\right)$$

$$+ Ec\left(\frac{\partial u}{\partial r}\right)^2 + \alpha.Ec.\rho_p\left(u - u_p\right)^2 \tag{7.13}$$

$$\rho_p\left[v_p\frac{\partial T_p}{\partial r} + u_p\frac{\partial T_p}{\partial z}\right] = \frac{\epsilon}{Pr}\left(\frac{\partial^2 T_p}{\partial r^2} + \frac{1}{r}\frac{\partial T_p}{\partial r}\right) - \rho_p\left(T_p - T\right) + \frac{3}{2}$$

$$+ Pr.\epsilon.Ec\frac{\partial}{\partial r}\left(u_p\frac{\partial u_p}{\partial r}\right) - \frac{3}{2}.Ec.Pr.\rho_p\left(u - u_p\right)^2 \tag{7.14}$$

The dimensionless boundary conditions to be satisfied are

$$z = 0, u = u_p = u_\infty, \rho_p = \rho_{p\infty}, T = T_\infty, T_p = T_\infty \tag{7.15}$$

$$r = 0, v_p = 0, \frac{\partial u_p}{\partial r} = 0, \frac{\partial \rho_p}{\partial r} = 0, v = 0, \frac{\partial u}{\partial r}, = 0, \frac{\partial T}{\partial r} = 0, \frac{\partial T_p}{\partial r} = 0 \tag{7.16}$$

$$r = \infty, u_p = 0, u = 0, T = 0, T_p = 0, \frac{\partial T}{\partial r} = 0, \frac{\partial u}{\partial r} = 0, \frac{\partial u_p}{\partial r} = 0, \frac{\partial T_p}{\partial r} = 0 \tag{7.17}$$

For nontrivial solutions, in addition to the boundary conditions, the following integral conditions should also be satisfied. Integral condition for nontrivial solutions of

$$\text{Or } \frac{d}{dz}\int_0^\infty ru^2\,dr = \alpha\int_0^\infty r\rho_p\left(u_p - u\right)dr \tag{7.18}$$

Integral condition for nontrivial solution of

$$\frac{d}{dz}\int_0^\infty ruT\,dr = E_c\int_0^\infty r\left(\frac{\partial u}{\partial r}\right)^2 dr + \alpha E_c\int_0^\infty r\rho_p\left(u-u_p\right)^2 dr + \frac{2\alpha}{3Pr}\int_0^\infty r\rho_p\left(T_p-T\right)dr \quad (7.19)$$

7.3   ANALYSIS OF THE THERMAL BOUNDARY LAYER

Taking $T = T_0 + T_1$ and $T_p = T_{p0} + T_{p1}$, where T_1 and T_{p1} are perturbation quantities, we get two sets of equations, viz. first set:

$$v_0\frac{\partial T_o}{\partial r} + u_0\frac{\partial T_o}{\partial z} = \frac{2\alpha}{3Pr}\rho_{p0}\left(T_{p0}-T_0\right) + \frac{1}{Pr}\left[\frac{\partial^2 T_0}{\partial r^2} + \frac{1}{r}\frac{\partial T_o}{\partial r}\right]$$

$$+ Ec\left(\frac{\partial u_o}{\partial r}\right)^2 + \alpha.Ec.\rho_{p0}\left(u_0-u_{p0}\right)^2 \quad (7.20)$$

$$\rho_{p0}\left(v_{po}\frac{\partial T_{po}}{\partial r} + u_{po}\frac{\partial T_{po}}{\partial z}\right) = \frac{\epsilon}{Pr}\left[\frac{\partial^2 T_{po}}{\partial r^2} + \frac{1}{r}\left(\frac{\partial T_{po}}{\partial r}\right)\right] - \rho_{po}\left[T_{p0}-T_0\right]$$

$$+ \frac{3}{2}Pr.\epsilon.Ec\frac{\partial}{\partial r}\left(u_{po}\frac{\partial u_{po}}{\partial r}\right) - \frac{3}{2}.Ec.Pr.\rho_{p0}\left(u_o-u_{p0}\right)^2 \quad (7.21)$$

Subject to the boundary conditions

$$r = 0, \frac{\partial T_o}{\partial r} = 0, \frac{\partial T_{po}}{\partial r} = 0 \quad (7.22)$$

$$r = \infty, T_0 = 0, \frac{\partial T_o}{\partial r} = 0, T_{po} = 0, \frac{\partial T_{po}}{\partial r} = 0 \quad (7.23)$$

and the following initial condition must be satisfied

$$\frac{d}{dz}\int_0^\infty ru_0T_0\,dr = Ec\int_0^\infty \left[\left(\frac{\partial u_0}{\partial r}\right)\right]^2 dr + \alpha Ec\int_0^\infty \rho_{p0}\left(u_0-u_{p0}\right)^2 dr$$

$$+ \frac{2\alpha}{3pr}\int_0^\infty r\rho_{p0}\left(T_{p0}-T_0\right)dr \quad (7.24)$$

Since it is assumed that T_0 is not affected by the presence of particles and the first and fourth terms in the right hand side of Equation (7.20) have no significant effect on T_0, the second and third terms in right hand side of Equation (7.24) are also dropped.

It will be convenient to represent the solution of Equation (7.20) by the superposition of two solutions of the form

$$T_0 = T_{00} + T_{01} \tag{7.25}$$

where T_{00} is the general solution of Equation (7.38) in the absence of the term $E_c \left(\dfrac{\partial u_0}{\partial r} \right)^2$, which is the term arising from frictional heat. Therefore, T_{00} satisfies the equation:

$$v_0 \frac{\partial T_{00}}{\partial r} + u_0 \frac{\partial T_{00}}{\partial z} = \frac{1}{Pr} \left(\frac{\partial^2 T_{00}}{\partial r^2} + \frac{1}{r} \frac{\partial T_{00}}{\partial r} \right) \tag{7.26}$$

and the boundary condition

$$r = 0, \frac{\partial T_{00}}{\partial r} = 0, r = \infty, T_{00} = 0 \tag{7.27}$$

Further the integral condition becomes $\dfrac{d}{dz} \displaystyle\int_0^\infty r\, u_o T_\infty dr = 0,$

$$\text{Or } 2\pi \int_0^\infty r\, u_0 T_\infty\, dr = \text{constant} = \frac{N_0}{T_\infty} (\text{say}) \tag{7.28}$$

Let

$$T_{00} = \frac{N_o}{T} \cdot \frac{1}{z} h(\eta) \tag{7.29}$$

Then Equation (7.26) reduces to

$$\eta h'' + h' + Pr(gh' + g'h) = 0 \tag{7.30}$$

The boundary condition becomes

$$\eta = 0; h' = 0, \eta = \infty, h = 0 \tag{7.31}$$

And the integral condition becomes

$$\int_0^\infty g'h\, d\eta = \frac{1}{2\pi} \tag{7.32}$$

Integrating Equation (7.19) once and using the boundary condition $\eta = 0; h' = 0, g = 0,$ we find $\eta h' + Pr\, gh = 0$

$$\text{Or } \frac{h'}{h} = -Pr. \frac{C^2\eta}{1 + \frac{C^2\eta^2}{4}} \tag{7.33}$$

Which on integration gives

$$h(\eta) = A \left(1 + \frac{C^2\eta^2}{2} \right)^{-2Pr} \tag{7.34}$$

Cleanly, for $\eta = \infty, h = 0$.

Hence, the constant A will be

$$A = \frac{1 + 2Pr}{8\pi} \text{ and } T_{00} = \frac{N_0}{T_\infty} \cdot \frac{1}{z} h(\eta)$$

$$\text{Or } T_{00} = \frac{N_0}{T_\infty} \cdot \frac{1}{z} \cdot \frac{1 + 2Pr}{8\pi} \left(1 + \frac{C^2\eta^2}{4} \right)^{-2Pr} \tag{7.35}$$

Further, T_{01} is a solution of the differential equation

$$v_0 \frac{\partial T_{01}}{\partial r} + u_0 \frac{\partial T_{01}}{\partial z} \frac{1}{Pr} \left(\frac{\partial^2 T_{01}}{\partial r^2} + \frac{1}{r} \frac{\partial T_{01}}{\partial r} \right) + E_c \left(\frac{\partial u_o}{\partial r} \right)^2 \tag{7.36}$$

With boundary condition, $r = 0, \dfrac{\partial T_{01}}{\partial r} = 0, r = \infty, T_{01} = 0$

The integral condition

$$\frac{d}{dz} \int_0^\infty r \, u_0 T_{01} \, dr = EC \int_0^\infty r \left(\frac{\partial u_0}{\partial r} \right)^2 dr \tag{7.37}$$

is identically satisfied

$$\text{For } Pr = 1, T_{01} = \frac{-Ecu^2_0}{2} \text{ is a solution of Equation (7.37)} \tag{7.38}$$

For arbitrary values of Pr

Let

$$T_{01} = \frac{-16c^2 EC}{z^2} H(\zeta) \tag{7.39}$$

Then, Equation (7.36) becomes

$$\xi H'' + H' + Pr \left(gH' + 2Hg' \right) - \frac{Pr}{16\xi} \left(g'' - \frac{g}{\xi} \right)^2 = 0 \tag{7.40}$$

Let $s = \dfrac{g(\xi)}{4}$

Then, Equation (7.40) reduces to

$$s(1-s)\frac{d^2H}{ds^2} + \{1-(1-\mathrm{Pr})2s\}\frac{dH}{ds} + 4\sigma H = \sigma s(1-s)^3 \tag{7.41}$$

With the boundary conditions

$$s = 0,\, H \text{ is finite},\, s = 1,\, H = 0 \tag{7.42}$$

The solution of Equation (7.41) subject to Equation (7.42) is given by

$$H = A_0\, 2F_1(\alpha,\beta,1,s) + H_P \tag{7.43}$$

where A_0 is an arbitrary constant and h_p is the particular integral. For any value of, except (1/3), (3/4), (5/3), (6/5),

$$H_p = As^4 + Bs^3 + Cs^2 + Ds + E \tag{7.44}$$

where

$$A = \frac{Pr}{4(5-3\,\mathrm{Pr})}, B = -\frac{Pr(11-9_{Pr})}{2(5-3_{Pr})(6-5_{Pr})}, c = \frac{3Pr(30_{Pr}^2 - 59_{Pr} + 27)}{4(3-4_{Pr})(6-5_{Pr})(5-3_{Pr})},$$

$$D = \frac{4c - Pr}{2(1-3Pr)} \text{ and } E = \frac{D}{4\,Pr},$$

$$A_0 = -(A+B+C+D+E)\frac{1}{2f_1(\alpha,\beta,1,1)}$$

Second set

$$\left[v_0\frac{\partial r_1}{\partial r} + v_1\frac{\partial T}{\partial r} + u_0\frac{\partial T_1}{\partial z} + u_1\frac{\partial T_o}{\partial}\right] = \frac{2\alpha}{3pr}\left(\rho_{po}Tp_1 - \rho_{po}T_1 + \rho_{p1}T_{po} - \rho_{p1}T_0\right)$$

$$+ \frac{1}{Pr}\left[\frac{\partial^2 T_1}{\partial r^2} + \frac{1}{r}\frac{\partial T_1}{\partial r}\right] + Ec\left(\frac{\partial u_1}{\partial r}\right)^2$$

$$+ \alpha.Ec.\left[2\rho_{po}\left(u_0 - u_{po}\right)\left(u_1 - u_{p1}\right) + \rho_{p1}\left(u_0 - u_{po}\right)^2\right]$$

$$\tag{7.45}$$

$$\rho_{p0}v_{p0}\frac{\partial T_{p1}}{\partial r} + \rho_{p0}v p_1\frac{\partial T_{p1}}{\partial r} + \rho_{p0}u_{p0}\frac{\partial T_{p1}}{\partial z} + \rho_{p0}u_{p1}\frac{\partial T_{p0}}{\partial z} + \rho_{p1}v_{p0}\frac{\partial T_{p0}}{\partial r} + \rho_{p1}u_{p0}\frac{\partial T_{p0}}{\partial z}$$

$$= \frac{\epsilon}{Pr}\left[\frac{\partial^2 T_{p1}}{\partial r^2} + \frac{1}{r}\left(\frac{\partial T_{p1}}{\partial r}\right)\right] - \left[\rho_{p0}T_{p1} - \rho_{p0}T_1 + \rho_{p1}T_{p0} - \rho_{p1}T_0\right]$$

$$+ \frac{3}{2}Pr.\epsilon.Ec\frac{\partial}{\partial r}\left(u_{p0}\frac{\partial u_{p1}}{\partial r} + u_{p1}\frac{\partial u_{p0}}{\partial r}\right)$$

$$- \frac{3}{2}.Ec.pr.\left[2\rho_{p0}\left(u_0 - u_{p0}\right)\left(u_1 - u_{p1}\right) + \rho_{p1}\left(u_0 - u_{p0}\right)^2\right] \tag{7.46}$$

Further, the solutions for T_1 and T_{p1} subject to the boundary conditions

$$r = 0: \frac{\partial T_1}{\partial r} = 0, \frac{\partial T_{p1}}{\partial r} = 0 \tag{7.47}$$

$$r = \infty: T_1 = 0, T_{p1} = 0 \tag{7.48}$$

are obtained by finite difference technique.

The integral condition

$$\frac{d}{dz}\int_0^\infty r(u_o T_1 + u_1 T_o)dr = Ec\int_0^\infty\left[2\left(\frac{\partial u_0}{\partial r}\right)\left(\frac{\partial u_1}{\partial r}\right)\right]dr$$

$$+ \alpha Ec\int_0^\infty 2\rho_{p0}\left(u_0 - u_{p0}\right)\left(u_1 - u_{p1}\right) + \rho_{p1}\left(u_0 - u_{p0}\right)^2 dr$$

$$+ \frac{2\alpha}{3Pr}\int_0^\infty\left(\rho_{p0}T_{p1} - \rho_{p0}T_1 - \rho_{p1}T_{p0}\right)dr \tag{7.49}$$

is identically satisfied.

7.4 DISCUSSION

Numerical computations have been made by taking $\epsilon = 0.05, 0.1 \& 0.2$; $Pr = 0.71$, 1.0 & 7.0; $\varphi = 0.001, 0.0001$; $\alpha = 0.1, 0.2$, $D = 0.5, 50, 100\,\mu m$, $EC = 0.0, 0.1$. The initial velocity and temperature profiles have been described at $z = 1.1$. The solution is obtained for $z = 1.2$ onwards with a step length of $\Delta z = 0.1$. From a number of numerical experiments, we have arrived at a conclusion that our implicit scheme is stable for any value of Δz and Δr.

Here, we have used a variable grid in the r-direction, where the grid growth ratio $r_r = \left(r_{j+1} - r_j\right)/\left(r_j - r_{j-1}\right)$. The resulting difference equations have been repeated $J_{max} - 2$ at the interior nodes forming a tridiagonal system of equations that can be solved using the Thomas algorithm to get the velocity, temperature, and the concentration profiles at the grid point $z + \Delta z$. The continuity equation is integrated across the boundary layer to give velocity v and for grid points at $z + \Delta z$.

From Figures 7.2 and 7.3, we observe that T_{p0} decreases with the increase of ϵ, the diffusion parameter without and with viscous heating, respectively. Similarly, from Figures 7.4 and 7.5, it can be concluded that T_{p1} increases with the increase of ϵ either without viscous heating or with viscous heating. This is perhaps due to the increase of resistive force which leads to generation of heat and heat is transferred from the particle to the fluid in the mixing region of the jet.

From Figure 7.6, it is concluded that T_{p0} decreases with φ for both viscous heating and without viscous heating. Figures 7.7 and 7.8 depict the temperature distribution T_{p1} with φ. T_{p1} increases with the increase of φ either in case of viscous heating or without viscous heating as it is due to increase of surface area of particles and the resistive force. In Figure 7.9, the mean and perturbed temperature profiles for particles displayed for different Prandtl numbers $Pr = 0.71$, 1.0, and 7.0 correspond to air,

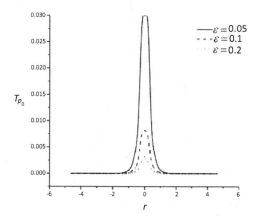

FIGURE 7.2   Variation of T_{p0} with r.

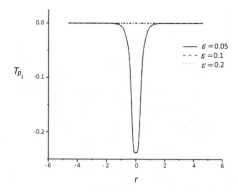

FIGURE 7.3   Variation of T_{p0} with r.

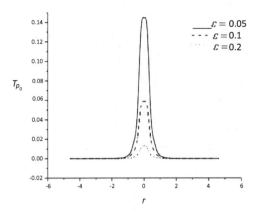

FIGURE 7.4 Variation of T_{p1} with r.

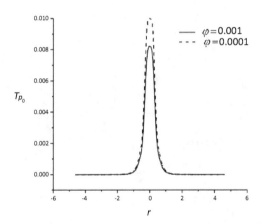

FIGURE 7.5 Variation of T_{p0} with r.

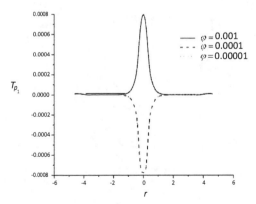

FIGURE 7.6 Variation of T_{p0} with r.

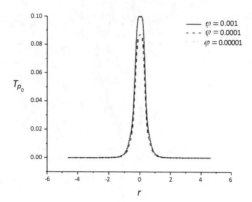

FIGURE 7.7   Variation of T_{p1} with r.

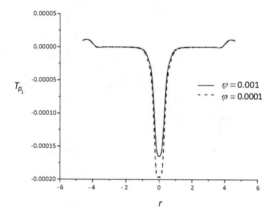

FIGURE 7.8   Variation of T_{p1} with r.

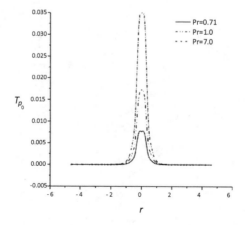

FIGURE 7.9   Variation of T_{p0} with r.

electrolyte solution, and water, respectively. It is observed that the temperature at the center of the jet is maximum in all the cases and decreases asymptotically to attain the free stream value. Further, the particle temperature increases with decrease in the conductivity of the carrier fluid at the center of the jet, as the transfer of heat from particle to fluid with less conductivity is always restricted than that of the fluid with more conductivity.

From Figure 7.10, we observe that T_{p1} increases with Pr when viscous heating is considered.

From Figure 7.11, we can observe that the temperature goes on decreasing at the center of the jet with the increase of z. Further from Figures 7.12 and 7.13, it can be concluded that the temperature of carrier fluid phase is greater than the particle phase.

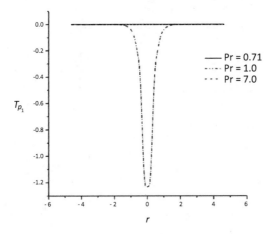

FIGURE 7.10   Variation of T_{p1} with r.

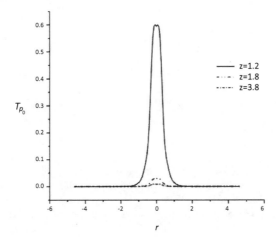

FIGURE 7.11   Variation of T_{p0} with r for different z.

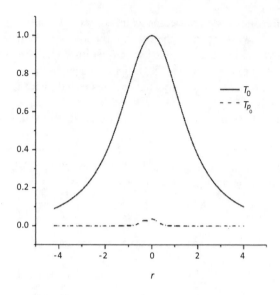

FIGURE 7.12   Variation of T_0 and T_{p0} with r.

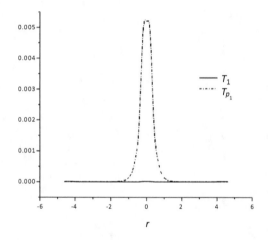

FIGURE 7.13   Variation of T_1 and T_{p1} with r.

REFERENCES

1. M. Fairweather and J.-P. Hurn. Validation of an anisotropic model of turbulent flows containing dispersed solid particles applied to gas–solid jets, *Computers & Chemical Engineering*, 32, 590–599, 2008.
2. J. J. L. Bansal. Jets of conductive fluids in the presence of a transverse magnetic field, *ZAMP*, 55, 479–489, 1975.
3. A. J. Chamkha. Hydromagnetic two-phase flow in a channel, *International Journal of Engineering Science*, 33, 437–446, 1995.

4. I. L. Ryhming. Zur ebenen laminaren zweiphasigen Strahlstrmung, *ActaMechanica*, 11, 117–140, 1971.

5. G. Rudinger, Dynamics of gas-particle mixtures with finite particle volume, *2nd Aerospace Sciences Meeting*, 1965.

6. H. Schlichting. *Boundary Layer Theory*, 7th edition. McGraw-Hill, New York, pp. 578–583, 1968.

7. A. Pozzi and A. Binachini, Linerized solutions for plane jets, *ZAMM*, 52, 523–528, 1972.

8. A. Pozzi and A. Binachini. Linerized solutions for circular jets, *ZAMM*, 54, 621–625, 1974.

9. S. L. Soo. *Fluid Dynamics of Multiphase Systems*. Blaisdell Publishing co., Waltham, MA, 1967.

10. N. Dutta and S. K. Das. Note on axis-symmetric jet mixing of compressible dusty fluid, *Acta Mechanica*, 86, 103–109, 1991.

11. N. Dutta and S. K. Das. Axially symmetrical jet mixing of an incompressible dusty fluid, *Acta Mechanica*, 55, 111–122, 1985.

8 A Review on Nearest-Neighbor and Support Vector Machine Algorithms and Its Applications

Ritesh Dash
Christain College of Engineering & Technology

Sarat Chandra Swain
KIIT University

CONTENTS

8.1 INTRODUCTION

Out of the different types of machine learning algorithms, data mining is the most common and significant. It is worth to mention here that people often make mistakes while predicting and analyzing the relationship between the variables and their dependent quantities. This is because there is an uncertain relationship between different quantities and their solutions. Machine learning can be applied to this type of problem for finding and predicting the accuracy and the relationship present between the datasets. Stochastic method of predicting using machine learning approach not only increases the efficiency of the system but also increases the accuracy level thereby minimizing the time consumption in predicting the performance.

A subset containing the features of datasets is usually prepared either as a training dataset or testing dataset. This feature may be either binary sets of data or a set of linguistic variables. Based on the above discussion, machine learning can be classified into three groups such as supervised learning, unsupervised learning, and reinforcement learning system. Supervised learning system uses a labeled a dataset, whereas unsupervised learning does not have labeled dataset. In contrast, reinforcement learning uses the output result as a feedback to train the system to predict in a more accurate and efficient manner. Semi-supervised learning system is another category of predicting the performance of the system which includes the characterization of both supervised and unsupervised learning systems.

The training dataset or the information is provided to the machine learning algorithm either in scalar or matrix form. The machine learning algorithm tries to create a relationship among different types of data and also measures the relationship among them so that during testing it could be able to find the best result out of the different possible values.

Out of different machine learning algorithms, this chapter is dedicated to some basic types of learning algorithms that find their applications in many basic problems. In this chapter, different types of classification problems and their types have been investigated from the literature.

The discussion is limited to a number of classification problems so as to achieve the goal of providing concrete and basic idea about the different types of machine learning algorithms and supervised machine learning algorithms. In the next section, we will discuss different types of issues present in supervised machine learning algorithm and their features while selecting particular algorithm to solve a particular kind of engineering problem. Section 8.3 provides different types of machine learning algorithms involved in predicting and analyzing some engineering problems. Section 8.4 gives a comparative analysis of all the machine learning techniques.

8.2 PROBLEM FACED BY SUPERVISED MACHINE LEARNING ALGORITHM

Machine learning approach is the process of analyzing and predicting an algorithm best suited for predicting a new set of data consisting of a subset of data or feature of the main algorithm. The process of applying supervised machine learning algorithm technique is as follows. The learning technique starts with collecting similar set of data and identifying required set of data. This required set of data is free from redundancy and error in data machine. Data preprocessing is technique which will identify all the features stated above. After data preprocessing, a training subset containing the features of main dataset is identified along with a particular algorithm which will best suit to the identified data. After successful completion of training parameters such as timestamp, epoch label, and error like market research Society, variance will identify whether training has been successfully conducted or not. PSEUDOCODE will be developed from the algorithm itself.

Selecting a particular algorithm is the most critical task. One has to test the data in terms of statistical analysis through SPSS, t-test, and chi-square test. The preliminary testing would gauge all the datasets, and, if possible, mapping would be applied for successful classification of data by applying a label to that dataset. The classifier's prediction most often depends upon prediction accuracy parameters, like the percent of correct predictions and the false predictions out of the total number of prediction datasets. However, during classification a split training dataset containing a training dataset consisting of particularly exclusive and equal-sized dataset is identified. Hence average error is found in the classifier technique. Out of the different types of cross-validation techniques, leave-1-cut technique is the most suitable one. Although this technique is most expensive, it can be used when more accurate dataset is required.

If the error rate is high during the classifier process, then we have to return to the previous classification algorithm and have to check if any one or all of the particular discrepancies are present, for example, relevant features of dataset were not taken into consideration while designing algorithm, there is requirement of a more number of datasets, dimensionally datasets are not balanced, the selected algorithm may not be appropriate for this particular problem, or tuning of more parameters is needed. After checking all those features if similar kind of problem persists then it may be concluded that datasets are not dimensionally correct or are imbalanced.

A common method of comparing different types of machine learning algorithms is to perform same kind of statistical analysis or comparison technique for checking their accuracy on different types of datasets.

If the data samples are of sufficient number than a training size of N, variables are tested on two different algorithms and the difference between the error and the accuracy of classifier must be studied. The difference between the two sets of data and their average in generalizing error across different training sets and their variance are calculated. To find out the variance among the two different techniques, t-test can be applied to check null hypothesis, that is, the mean of the two different classifiers must be zero. t-Test will produce two types of errors: type1, which rejects the null hypothesis; and type 2, in which null hypothesis may not be rejected in terms of its probability. If type 1 error is achieved, then the corresponding error will be close to the considered dataset.

Supervise learning algorithm is one of the most critical tasks that is frequently used with algorithm system for logical symbolical forecasting or classification technology. In the next section, a review of different types of supervise learning algorithm and their characteristics is presented.

8.3 NEAREST NEIGHBOR

In pattern recognition, nearest-neighbor technique and k-means technique are the most commonly used techniques for both classification and regression analysis. It usually consists of a k-closet set training dataset with feature containing main dataset or feature space. The output depends on the purpose for which it is applied. During classification analysis, the output consists of a single object-oriented value where k is an integer of a very small value of magnitude, which increases during classification with the objective to assign the shortest and closet value to the object to be optimized. While in regression analysis, the output is the average of k nearest neighbors. The conversion of this technique is slower because of the property of the function to approach the local value until and unless classification is properly carried out. For both classification and regression analysis, the intelligent technique is to assign a single value weight to continuous neighbors such that the nearest neighbor contributes more to the average of the value. Again, for classification purpose, a set of object properties from the parent dataset, whose characteristics and logical relationship are well known, are chosen.

Let

$$(x_1, y_1), (x_2, y_2), (x_3, y_3)...(x_n,)$$

where *[1,2]

where y represents the level of x, such that $y = r$.

For k nearest-neighbor technique, the best choice for the value of k depends upon the number of data, for example, longer dataset reduces the level of noise during the classification process. But increasing the dataset creates a boundary value between the two classes of data. Therefore, selection of k clearly depends upon the heuristic technique. However, there are some special cases where some special classification is predicted to the closet training set or sample called nearest-neighbor technique.

Let Cn^w denote the weighted nearest classifier with weight subject to regularity condition on the class distribution, then the excess risk has the following asymptotic expansion:

$$R_R\left(C_n^{wnn}\right) - R_R\left(C^{Bayes}\right) = \left\{B_1 S_n^2\right\}\{1+0\}$$

For constants

$$B_1 \text{ and } B_2 \text{ where } = \sum_{i=1}^n w \text{ and}$$

$$t_n = n^{-\frac{2}{d}} \sum_{i=1}^n w_{ni} \left\{ i^{1+\frac{2}{d}} - (i-1)^{1+\frac{2}{d}} \right.$$

The optimal weighting scheme $\left\{w_{ni}^*\right\}_i^n$, which balances the two terms in the above equation, is given as follows:

Set $K^* = \left[B_n^{\frac{4}{d}} \right.$

$$w_{ni}^* = \frac{1}{k^*}\left[1 + \frac{d}{2} - \frac{d}{2k^{*\frac{2}{d}}} \left\{ i^{1+\frac{2}{d}} - (i-1)^{1+\frac{2}{d}} \right\} \right]$$

For $i = 1, 2...$ and

$$w_{ni}^* = 0 \text{ for } i = k^* + 1...n$$

With optimal weights, the dominant term in the asymptotic expansion of excess risk is $o\left(n^{-\frac{d}{d}} \right)$. Similar results are true when using a bagged nearest-neighbor classifier.

Algorithm steps for k-means clustering:

Let $X = \{x_1, x_2, x_3, \ldots x_n\}$ be the set of data points, and $= \{v_1, v_2, \ldots\}$ be the set of centers.

   i. Randomly select c cluster center.
   ii. Calculate the distance between each data point and cluster center.
   iii. Assign the data point to the cluster center whose distance from the cluster center is minimum of all the cluster centers.
   iv. Recalculate the new cluster center using:

$$v_i = \left(\frac{1}{c_i}\right)\sum_{j=1}^{c_i} x_i$$

      where c_i represents the number of data points in i cluster.
   v. Recalculate the distance between each data point and new obtained cluster centers.
   vi. If the number of data point was reassigned, then stop otherwise repeat from step iii.

8.4 SUPPORT VECTOR MACHINE

Support vector machine can be classified into two types, namely, support vector classification and support vector regression. Support vector classification is again classified into three types as liner separable case (hard margin), linear nonseparable case (soft margin), and nonlinear case (kernel machine). Similarly, support vector regression is of two types, namely, support vector regression and nonlinear support vector regression. Support vector machine may be defined as an algorithm to find linear separability hyperplane of a binary labeled dataset. It is different from perception learning algorithm (PLA) in assigning and quantifying a hyperplane. Support vector machine always requires a separability hyperplane that can maximize the margin between the positive and negative sample. So, from different support vector machines the objective function for finding out the relationship between different types of data can be written as

$$\max_{w,b} \min_{w,b} y^n \left(w^t x + b \parallel w\right)$$

Here, w represents the distance between the two clusters of centroid hyperplane, and b represents the difference center hyperplane to any one inside of the cluster dataset.

The objective function as shown above, it can be concluded that this is an constraint problem where the objective is to finalize

$$st.y^n \left(w^\tau u^n + b\right)_1$$

The main issues of support vector machine are the model flexibility and computational complexity. Computational complexity represents successful separation of the datasets.

The idea of support vector machine is to create a hyperplane, which indicates a particular class of data to which it belongs, to determine the largest distance to the nearest training datasets.

8.5 CONCLUSION

The main issues of support vector machine are the model flexibility and computational complexity. Computational complexity represents successful separation of the datasets. The idea of support vector machine is to create a hyperplane, which indicates a particular class of data to which it belongs, to determine the largest distance to the nearest training datasets.

REFERENCES

1. B. Agarwal, N. Mittal (2014). Text classification using machine learning methods-A survey. A.V. Babu, A. Nagar, K. Deep, M. Pant, J.C. Bansal, K. Ray, U. Gupta (eds.) *Proceedings of the Second International Conference on Soft Computing for Problem Solving*, December 2014, Jaipur, India.
2. E. Alpaydin (2014). *Introduction to Machine Learning*. MIT Press, Cambridge, MA.
3. B. Kitchenham, S. Charters (2007). Guidelines for performing systematic literature reviews in software engineering. Technical report, Ver. 2.3, EBSE, UK.
4. K.P. Bennett, E.J. Bredensteiner (2000). Duality and geometry in SVM classifiers. *Proceedings of the Seventeenth International Conference on Machine Learning*, Francisco, USA.
5. D.M. Blei, A.Y. Ng, M.I. Jordan (2003). Latent dirichlet allocation. *Journal of Machine Learning Research*, 3, pp. 993–1022, Jordan.
6. J. Carletta (1996). Assessing agreement on classification tasks: The kappa statistic. *Computational Linguistics*, 22(2), pp. 249–254.
7. A. Casamayor, D. Godoy, M. Campo (2010). Identification of non-functional requirements in textual specifications: A semi-supervised learning approach. *Information and Software Technology*, 52(4), pp. 436–445.

9 Eigenvalue Assignment for Control of Time-Delay System
A Lambert W Function–Based Approach

Jayanta Kumar Kar and Sovit Kumar Pradhan
National Institute of Technology

CONTENTS

9.1   INTRODUCTION

Most of the physical systems have time delay and are inherently nonlinear in nature. Time-delay systems (TDS) arise from inherent time delay in the components of the systems, or from the deliberate introduction of time delay into the systems for control purposes. TDSs can be represented by delay differential equation (DDE), which belong to the class of functional differential equation and have been studied over the past decades. Thus, it becomes essential to incorporate the effect of delay and nonlinearity in the stability analysis.

The presence of delay and nonlinearity in a system complicates the stability and performance analysis. Time delay can be constant or time-varying.

Time delays are usually encountered in numerous technological systems, like electrical, pneumatic, and hydraulic networks; chemical processes; communication networks; transmission lines; robotics; etc. Existence of time delay, regardless of its presence within the control or/and the state, might cause undesirable system transient response, or even instability. Subsequently, the problem of observability, controllability, robustness, pole placement, adaptive control, and stabilization of systems for this category has been the most interesting area of research throughout the past few decades.

9.2 LAMBERT *W* FUNCTION

9.2.1 PRELIMINARY

Lambert *W* function (LWF), also known as omega function, was introduced in the 1700s by Lambert and Euler. The source of the LWF is Lambert's transcendental equation, derived by Lambert in 1758. However, his research did not become well known until Euler wrote about it in his paper in the year 1783 where a special case of the LWF, which reduces to $Wa^w lx$, was presented. The *W* part of this function's name comes from Polya and Szego who used the *W* symbol to represent the Lambert function in their research in 1925. In 1980, the LWF came into more popular recognition when it was utilized by the Maple computer algebra system. It was within the Maple programming that the name Lambert *W* was introduced for this function [1–3].

9.2.2 MATHEMATICAL FORM OF LAMBERT *W* FUNCTION

LWF can be defined for any function $W(x)$, satisfying the below relationship:

$$W(x)e^{W(x)} = x$$

The LWF is a function which has a complex value, with a complex argument, x, having an infinite number of branches, W_k, where $k = -\infty, \ldots, -1, 0, 1, \ldots, \infty$. The real part of the solution to the LWF changes with different values of x, where W represents the LWFs in terms of x.

9.3 SOLUTION OF DELAY DIFFERENTIAL EQUATIONS USING LAMBERT *W* FUNCTION

In this chapter, an analytical approach for the solution to the systems of DDE has been studied for homogeneous scalar and some special cases of systems of DDE using the LWF. The obtained solution is outlined in terms of an infinite series of modes written in terms of the matrix LWF. This solution has similarity to the concept of the state transition matrix in linear ordinary differential equation (ODE), enabling its use for general classes of linear DDE.

9.3.1 FREE SYSTEM OF DDEs: SCALAR CASE

An analytic approach for the complete solution of systems of DDE based on the notion of the LWF was developed by Sun Yi et al. They considered a first-order scalar homogeneous DDE as

$$\dot{x}(t) = ax(t) + a_d x(t-h), \ t > 0$$

$$x(t) = x_0, \ t = 0 \tag{9.1}$$

$$x(t) = g(t), \ t \in [-h, 0)$$

Here, for DDE two initial conditions are needed to be specified, instead of a simple initial condition as in ODE: a preshape function, $g(t)$ for $-h \leq t < 0$, and an initial point, at time $t = 0$. The parameter denotes the delay time. The solution can be derived using the LWF in terms of infinite number of branches of the LWF as

$$x(t) = \sum_{k=-\infty}^{\infty} e^{S_k t} c_k^I \tag{9.2}$$

$$S_k = \frac{1}{h} W_k \left(a_d h e^{-ah} \right) + \tag{9.3}$$

where the coefficient C_k^I can be determined numerically from the preshape function, $g(t)$, and an initial state, x_0, is defined in the Banach space. Unlike the results of other existing methods, the solution in Equation (9.2) has an analytical form expressed in terms of the parameters of the DDE in Equation (9.1), that is, a, a_d, and h. One can explicitly determine how the parameters are involved in the solution and how each parameter affects each eigenvalue. Also, each eigenvalue is differentiated by k, which indicates the branch of the LWF [12].

9.3.2 MATRIX CASE

A system of DDEs in matrix-vector form is defined as

$$\dot{x}(t) = Ax(t) + A_d x(t-h), \ t > 0$$

$$X(t) = x_0 \ t = 0 \tag{9.4}$$

$$X(t) = g(t) t \in [-h, 0)$$

where A and A_d are $n \times n$ matrices, and $x(t)$ is a $n \times 1$ state vector, and $g(t)$ are specified preshape functions and an initial state defined in the Banach space, respectively. Informally, a Banach space is a vector space with a metric that allows the computation of vector length and distance between vectors and is complete in the sense that a Cauchy sequence of vectors always converges to a well-defined limit in the space. The existence and uniqueness of the solution for this system of linear

DDE given in Equation (9.4) have been proved. A special case where the coefficient matrices A and A_d commute the solution of DDE is

$$x(t) = \sum_{k=-\infty}^{\infty} e^{\left(\frac{1}{h}W_k\left(A_d h e^{-Ah}\right)+A\right)t} C_k^I \tag{9.5}$$

However, this solution has the same form as for scalar case, which is valid when the matrices A and A_d commute, that is, $AA_d = A_d A$. Therefore, it cannot be used for general systems of DDE, which means this solution is not exact. Hence, the exact solution in terms of the LWF for systems of DDE (Equation 3.9) for the general case is given later. First, a solution form for Equation (9.9) is assumed as

$$x(t) = e^{St} C^I \tag{9.6}$$

where S is $n \times n$ matrix and C^I is constant $n \times 1$ vector. The characteristic equation for Equation (9.3) can be obtained by assuming a nontrivial solution of the form, where S is a scalar variable, and C^I is a constant. Also, one can assume the form of Equation (9.6) for deriving the solution to DDE using LWF. Substituting Equation (9.6) into (9.9) yields

$$Se^{St} C^I - Ae^{St} C^I - A_d e^{S(t-h)} C^I = 0 \tag{9.7}$$

$$Se^{St} C^I - Ae^{St} C^I - A_d e^{S(t-h)} C^I = \left(S - A - A_d e^{-sh}\right) e^{St} C^I = 0 \tag{9.8}$$

Because the matrix S is an inherent characteristic of a system, and independent of initial conditions, it can be concluded that Equation (9.8) be satisfied for any arbitrary initial condition and for every time t

$$S - A - A_d e^{-Sh} = 0 \tag{9.9}$$

In the special case where $A_d = 0$, the delay term in Equation (9.9) disappears, and it becomes a system of ODE, and Equation (9.9) reduces to

$$S - A = 0 \Leftrightarrow S = A \tag{9.10}$$

Substituting Equation (9.10) into (9.6), it becomes a system of ODEs only with x_0 without $g(t)$ (i.e., $C^I = x_0$), yielding

$$x(t) = e^{At} x_0 \tag{9.11}$$

This is the well-known solution to a homogeneous system of ODE in terms of the matrix exponential. Returning to the system of DDE in Equation (9.9), one can multiply $he^{Sh}e^{-Ah}$ on both sides of Equation (9.10) and rearrange to obtain

$$h(S - A)e^{Sh}e^{-Ah} = A_d h e^{-Ah} \tag{9.12}$$

Generally, matrices S and A do not commute. When A and A_d commute, then S and A_d also commute, and the relationship is

$$h(S-A)e^{Sh}e^{-Ah} \neq h(S-A)e^{(S-A)h} \tag{9.13}$$

Consequently, to compensate for the inequality in Equation (9.13) and to use the matrix LWF defined as

$$W(H)e^{W(H)} = H \tag{9.14}$$

Here, an unknown matrix Q is introduced to satisfy

$$h(S-A)e^{(S-A)h} = A_d hQ \tag{9.15}$$

On comparison of Equations (9.14) and (9.15)

$$(S-A)h = W(A_d hQ) \tag{9.16}$$

Then, the solution matrix, S, is obtained by solving Equation (9.16)

$$S = \frac{1}{h}W(A_d hQ) + A \tag{9.17}$$

Substituting Equation (9.17) into (9.12) yields the following condition which can be used to solve for the unknown matrix Q

$$W(A_d hQ)e^{W(A_d hQ)+Ah} = A_d h \tag{9.18}$$

The matrix LWF, $W(H)$, is complex valued, with a complex argument H, and has an infinite number of branches $W_k(H_k)$, where $k = -\infty, \ldots, -1,0,1,\ldots\infty$. Corresponding to each branch, k, of the LWF, W_k, there is a solution Q_k from Equation (9.18), and for $H_k = A_d hQ_k$, the Jordan canonical form J_k is computed from $H_k = Z_k J_k Z_k^{-1}.J_k = \text{diag}\left(J_{k1}\left(\hat{\lambda}_1\right), J_{k2}\left(\hat{\lambda}_2\right), \ldots, J_{kp}\left(\hat{\lambda}_p\right)\right)$, where $J_{ki}\left(\hat{\lambda}_i\right)$ is $m \times m$ Jordan block and m is multiplicity of the eigenvalue $\hat{\lambda}_i$. Then, the matrix LWF can be computed as

$$W_k(H_k) = Z_k \left\{ \text{diag}\left(W_k\left(J_{k1}\left(\hat{\lambda}_1\right)\right), \ldots, W_k\left(J_{kp}\left(\hat{\lambda}_p\right)\right)\right)\right\} \tag{9.19}$$

where

$$W_k\left(J_{ki}\left(\hat{\lambda}_i\right)\right) = \begin{bmatrix} W_k\left(\hat{\lambda}_i\right) & \cdots & \frac{1}{(m-1)!}W_k^{(m-1)}\left(\hat{\lambda}_i\right) \\ \vdots & \ddots & \vdots \\ 0 & \cdots & W_k\left(\hat{\lambda}_i\right) \end{bmatrix} \tag{9.20}$$

With the matrix LWF, W_k, given in Equation (9.19), S_k is computed from Equation (9.17). The principal $(k = 0)$ and other $(k \neq 0)$ branches of the LWF can be calculated from a series definition or using commands already embedded in various commercial software packages, such as MATLAB® and Maple. With W_k, this satisfies

$$W_k(H_k)e^{W_k(H_k)} = H_k \tag{9.21}$$

Finally, the matrix Q_k is obtained from

$$W(A_d h Q)e^{W(A_d h Q) + Ah} = A_d h \tag{9.22}$$

and the obtained matrix Q_k is substituted into Equation (9.17)

$$S_k = \frac{1}{h}W(A_d h Q) + A \tag{9.23}$$

$$x(t) = \sum_{k=-\infty}^{\infty} e^{S_k t} C_k^l \tag{9.24}$$

The coefficient C_k^l in Equation (9.24) is a function of A, A_d, h, and the preshape function, $g(t)$, and the initial state, x_0.

9.4 PROPORTIONAL-INTEGRAL CONTROL OF SYSTEMS WITH TIME DELAY

In this chapter, a new design method is presented for proportional-integral (PI) controllers of first-order plants in the presence of time delays. In general, time delays can limit and degrade the achievable performance of the controlled systems, and even induce instability. Thus, PI gains should be selected carefully considering such effects of time delays [11]. Unlike existing methods, the design method presented here is based on solutions to DDE, which are derived in terms of the LWF. PI controllers for first-order plants with time delays are designed by obtaining the rightmost eigenvalues in the infinite eigen spectrum of TDS and assigning them to desired positions in the complex plane [10]. The controllers designed using the proposed method can improve the system performance and successfully stabilize an unstable plant.

9.4.1 ILLUSTRATIVE EXAMPLE

A first-order plant with time delay can be described as

$$G(s) = G_p(s)e^{-sh} = \frac{K_M}{\tau_M s + 1}e^{-sh} \tag{9.25}$$

where τ_M is the time constant, h is the time delay, and K_M represents the steady-state gain.

Consider the PI controller

$$G_s = \overline{K_P} + \frac{\overline{K_I}}{s} \tag{9.26}$$

Many controllers in industrial processes only have PI action and such controllers are widely used, for example, in automotive controllers. A primary goal of this is to choose the gains K_P and K_I such that a stable closed loop system with improved transient response (i.e., reduced rise time) is obtained. Here, barred variables ($\overline{}$) denote gains selected by using the Smith predictor (SP) approach to distinguish them from gains selected by using the LWF approach. Where PI controllers are sufficient for all systems that have first-order transfer functions to obtain zero steady-state error and adequate transient responses and they are widely used for controlling numerous industrial processes [6,13]. However, it is well known that the longer the time delay, the more difficult it is to stabilize the system. Moreover, the delay term in the closed-loop characteristics equation complicates the stability analysis and the design of the controller to guarantee stability [13].

The SP results in a delayed response of a delay-free system by moving the time delay outside the feedback loop only when the model in the SP, that is, $\tilde{G}_P(s)$ and $\tilde{h}$, is assumed to be exactly the same as the plant, that is, $G_P(s)$ and h. Then, the controller $G_s(s)$ in Equation (9.2) can be designed considering only the delay-free plant $G_P(s)$ (Figure 9.2). This is the main advantage of the SP control. For example, in order to meet given time domain specifications, the desired eigenvalues can be chosen from the desired natural frequency ω_n and the desired damping ratio ζ, as

$$\lambda_d = -\omega_n \zeta \pm \omega_n \sqrt{1 - \zeta^2} i = -\sigma \pm \omega_d i$$

Assuming no time delay, $\overline{K_P}$ and $\overline{K_I}$ are chosen as

$$\overline{K_I} = \frac{\omega_n^2 \tau_M}{K_M}, \quad \overline{K_P} = \frac{2\zeta\omega_n \tau_M - 1}{K_M}$$

By substituting the desired values into the closed-loop characteristics equation of the system in Figure 9.2

$$s^2 + \left(\frac{1 + K_M \overline{K_P}}{\tau_M} \right) + \frac{K_M \overline{K_I}}{\tau_M} = 0 \tag{9.27}$$

This means that the controller $G_s(s)$ can be designed considering only the nondelayed part $\tilde{G}_P(s)$ of the plant ignoring the time delay e^{-sh}. This method is, however, based on pole-zero cancellation and, thus, the stability is vulnerable to uncertainty in system parameters. Careful modeling and parameter identification are crucial for its successful application. Furthermore, the function cannot handle disturbances and nonzero initial conditions. It is well known that SP-based controllers are sensitive to uncertainties in the delay. Various extensions and modifications to the SP have been proposed to address its limitations.

9.4.2 LAMBERT W FUNCTION–BASED APPROACH

In this section, as an alternative to the SP, a design approach for the PI controller is developed via rightmost eigenvalue assignment using the LWF [4,5]. The open-loop transfer function with a PI controller is

$$G_{open} = \frac{K_M}{\tau_M s + 1} e^{-sh} \left(K_P + \frac{K_I}{s} \right) = \frac{y}{e} \tag{9.28}$$

where $e = -y + r$. Then, the closed-loop system in the time domain becomes

$$\ddot{y} = -\frac{1}{\tau_M} \dot{y} - \frac{K_M K_I}{\tau_M} \dot{y}(t-h) - \frac{K_P K_I}{\tau_M} y(t-h) + \{ K_M K_P s + K_M K_I \} r(t-h)$$

By defining $x_1 = y$, $x_2 = \dot{y}$, Equation (9.7) can be rewritten as

$$\dot{x}_1 = x_2$$

$$\dot{x}_2 = -\frac{1}{\tau_M} x_2 - \frac{K_M K_P}{\tau_M} x_2(t-h) - \frac{K_M K_I}{\tau_M} x_1(t-h) + \{ K_M K_P s + K_M K_I \} r(t-h) \tag{9.29}$$

Then, we obtain the closed-loop system in state-space form, where the reference input is ignored to focus on stability as

$$\dot{X}(t) = \begin{bmatrix} 0 & 1 \\ 0 & -\dfrac{1}{\tau_M} \end{bmatrix} X(t) - \begin{bmatrix} 0 & 0 \\ \dfrac{K_M K_I}{\tau_M} & \dfrac{K_M K_I}{\tau_M} \end{bmatrix} \tag{9.30}$$

From the roots of characteristic Equation (9.30), the eigenvalues of the system are obtained. However, due to the time delay e^{-sh}, the system is infinite dimensional and, thus, there exist an infinite number of eigenvalues. The principal difficulty in analyzing and controlling systems with time delays arises from this transcendental character, and the determination of this eigen spectrum typically requires numerical, approximate, or other approaches. Obtaining and controlling the entire infinite eigen spectrum is not as straightforward as for systems of ODE. Instead, for DDE, such as Equation (9.30), it is desired to locate the dominant eigenvalues, which are rightmost in the complex plane, and to assign them to the desired positions. The LWF-based approach is an efficient tool for doing this, and in the subsequent section the approach is applied to Equation (9.30) to design the PI controller and compared with the SP. With the coefficient matrices A and A_d defined in Equation (9.30), the solution matrix S_0 is computed as

$$S_0 = \frac{1}{h} W_0 \left(A_d h Q_0 \right) + A \tag{9.31}$$

where the unknown matrix Q_0 is obtained by solving

$$W_0 \left(A_d h Q_0 \right) e^{W_0 (A_d h Q_0) + Ah} = A_d h$$

Equation (9.31) can be solved numerically to obtain the matrix Q_0 for the principal branch ($k = 0$) using nonlinear solvers. Then, by substituting the matrix Q_0 we obtain S_0 and its eigenvalues. Then, one can set the desired locations for eigenvalues equal to those of S_0, that is, $\lambda_i (S_0) = \lambda_{i,\,\text{desired}}$ for $i = 1,\dots,n$, where $\lambda_i (S_0)$ is the ith eigenvalue of the matrix S_0.

The gains K_P and K_I are then obtained by solving the equation numerically. The desired positions for the rightmost eigenvalues, that is, λ_i desired, can be chosen from the desired natural frequency and damping ratio using Equation (9.3). In the assignment of eigenvalues, the principal branch of the LWF is used to find the rightmost eigenvalues [3,8]. For the first-order scalar DDE, it has been proven that the rightmost eigenvalues are always obtained by using the principal branch. For general systems of DDE, such a proof is not available. It has been observed that, if the coefficient A_d does not have repeated zero eigenvalues, the rightmost eigenvalues are obtained with the principal branch. If A_d has repeated zero eigenvalues, they are obtained by using the branches $k = 0$ and $k = -1$.

Consider the unstable system

$$G(s) = G_P(s) e^{-sh} = \frac{1}{(s-1)} e^{-0.1s} \tag{9.32}$$

and the PI controller in Equation (9.2). For such unstable plants, tuning algorithms based on Ziegler–Nichols methods as considered and approximated analytical solutions can be used to obtain stabilizing PI gains. Also, PI controllers for such unstable plants with time delays have been designed on the basis of bifurcation methods. That is, purely imaginary variables are substituted into the characteristic Equation (9.25) for the characteristic root. Then, the characteristic equation is divided into an imaginary and a real part. Solving the two equations simultaneously using numerical nonlinear solvers yields the ranges for K_P and K_I that stabilize the unstable plants as in Equation (9.32). Such bifurcation methods are combined with the Hermite–Biehler theorem, which is a theorem on stability of quasi-polynomials, and the regions of $K_I - K_P$ are derived. The PI controllers can be designed by using the small gain theorem and H∞ norms of transfer functions [7, 9]. On the other hand, stabilizing gains can also be obtained by assigning the positions of the poles of the closed-loop system. Because many properties of a closed-loop system depend on the location of its poles, pole placement is one of the mainstream methods in control system design. Pole assignment is especially useful when time-domain specifications need to be considered (e.g., rise time, maximum overshoot, etc.). If it is not feasible to assign all poles, the behavior of a system can often be characterized with a few dominant poles. Thus, one can attempt to place a few dominant poles. The SP is a well-known method for pole placement.

9.5   RESULT ANALYSIS

The gains of the PI controller are found to be $\bar{K}_I = 3.0625$ and $\bar{K}_P = 2.8900$, when the desired natural frequency is $\omega_n = 1.75$ and the desired damping ratio is $\zeta = 0.59$. With those gains, one can assign the eigenvalues of the system to be $-0.9592 \pm 1.9696i$, which are stable theoretically. However, due to the initial conditions, disturbance, and errors in simulation, this control leads to instability. If Equation (9.27) is unstable, the characteristic equation of the closed-loop system with the SP in Figure 9.1 retains the unstable pole of the open-loop system. Therefore, the SP cannot stabilize the system. Figure 9.1 shows responses simulated using Simulink. Even though there is no disturbance or initial condition mismatch, due to errors in numerical integration the SP cannot successfully stabilize the unstable plant in Equation (9.12). On the other hand, the LWF-based approach does not reduce the number of eigenvalues. Instead, it assigns the rightmost eigenvalues of the infinite spectrum to the desired positions.

For example, the resulting gains for the rightmost eigenvalues $-0.9592 \pm 1.9696i$ ($\omega_n = 1.75$ and $\zeta = 0.59$) are $K_I = 2.2217$ and $K_P = 2.6983$.

Table 9.1 shows the PI gains corresponding to other values of ω_n. In designing PI control for systems of ODE, an increase in ω_n induces a decrease in rise time. As seen in Figure 9.1, by adjusting the values of ω_n, one can tune the gains to meet time-domain specifications in a similar way to ODE. Because the LWF-based approach does not require pole-zero cancellation, the designed controller safely stabilizes the system. Another prediction-based approach, that is, finite spectrum assignment, also has unsolved problems regarding integral approximation and, thus, can fail to stabilize unstable systems (Table 9.2).

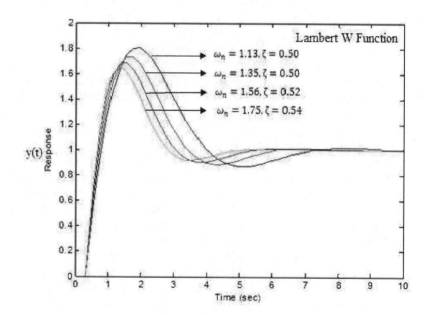

FIGURE 9.1   Simulated responses of the unstable system controlled with PI controllers designed using the LWF-based approach.

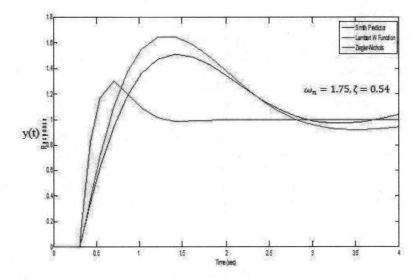

FIGURE 9.2 Responses of systems controlled using the SP, LWF, and Ziegler–Nichols method.

TABLE 9.1

Gains K_I and K_P of PI Controller Obtained by Using the LWF and SP

		Dominant Pole	Smith Predictor		Lambert W Function	
ω_n	ζ	$\lambda_d = -\zeta\omega_n \pm \omega_n\sqrt{1-\zeta^2}\,i$	$\bar{K}_I$	$\bar{K}_P$	K_I	K_P
1.13	0.50	$-0.6888 \pm 1.1724i$	1.2769	2.1300	1.0125	2.0136
1.35	0.50	$-0.5673 \pm 0.9776i$	1.8225	2.3500	1.4232	2.2238
1.56	0.52	$-0.8139 \pm 1.3365i$	2.4336	2.6224	1.8334	2.4367
1.75	0.54	$-0.9592 \pm 1.4696i$	3.0625	2.8900	2.2217	2.6483

TABLE 9.2

Transient Characteristics of Unstable Plant

ω_n	ζ	$\lambda_d = -\zeta\omega_n \pm \omega_n\sqrt{1-\zeta^2}\,i$ (Dominant Pole)	Rise Time (s)	Settling Time (s)	Maximum Overshoot (%)	Peak Time (s)
1.13	0.50	$-0.6888 \pm 1.1724i$	0.477	7.34	69.4	1.82
1.35	0.50	$-0.5673 \pm 0.9776i$	0.437	6.20	61.6	1.54
1.56	0.52	$-0.8139 \pm 1.3365i$	0.404	5.47	56.0	1.36
1.75	0.54	$-0.9592 \pm 1.4696i$	0.380	4.95	51.8	1.27

9.6 CONCLUSION

In this approach, the analysis of TDS, based on the LWF, has been done. Analytical solutions to systems of DDE are derived with stability determined on the basis of the eigenvalues of the principal branch of the LWF, that is, $k = 0$. The LWF-based method eliminates the need to add traditional time delay approximations to represent delays in the system model, and hence makes the solution for DDE a much easier process. A controller for time delay processes is designed based on the eigenvalue assignment algorithm using LWF and implemented on unstable plant. Simulation results have been compared with other existing methods. Performance indices measures are calculated to verify the controller response. The controller design using LWF is found to perform better as compared with other methods in terms of system characteristics, that is, steady-state error, rise time, overshoot, etc.

REFERENCES

1. Yi S., Ulsoy A. G., Nelson P.W., The Lambert W Function Approach to Time Delay Systems and the Lambert W_DDE Toolbox. *Proceedings of the 10-th IFAC Workshop on Time-Delay Systems, The International Federation of Automatic Control*, North eastern University, Boston, USA, June, 2012.
2. Alfergani, A., et al. Maximum allowable delay bound estimation using Lambert W function. *2017 IEEE Jordan Conference on Applied Electrical Engineering and Computing Technologies (AEECT)*, Jordan. IEEE, 2017.
3. Yi S., Ulsoy A. G., Nelson P.W., Eigenvalue assignment via the Lambert W function for control of time-delay systems. *Journal of Vibration and Control*, 16, 961–982 (2010).
4. Yi S., Ulsoy A.G., Nelson P.W., Controllability and observability of systems of linear delay differential equations via the matrix Lambert W function. *IEEE Transactions on Automatic Control*, 53, 854–860 (2008).
5. Choudhary N., Sivaramakrishnan J., Kar I.N., A robust sliding mode control approach for uncertain delay systems using Lambert W function. *TENCON 2018-2018 IEEE Region 10 Conference*, South Korea. IEEE, 2018.
6. Cheng Y.-C., Hwang C., Use of the Lambert W Function for time-domain analysis of feedback fractional delay systems. *Control Theory and Applications, IEE Proceedings*, USA, 2006.
7. Yi S., Ulsoy A.G., Solution of a system of linear delay differential equations using the matrix Lambert W function, *Proceedings of the 2006 American Control Conference*, Minneapolis, Minnesota, USA, June, 2006.
8. Wang Z.H., Hu H.Y., Calculation of the rightmost characteristic root of retarded time-delay systems via Lambert W function, *Journal of Sound and Vibration*, 318, 757–767, 2008.
9. Chen Y.Q., Moore K.L., Analytical stability bound for delayed second-order systems with repeating poles using Lambert W function, *Automatica*, 38, 891–895, 2002.
10. Jarlebring E., Damm T., The Lambert W function and the spectrum of some multidimensional time-delay systems, *Automatica*, 43, 2124–2128, 2007.
11. Gerov, R., Jovanović, Z., Synthesis of the proportional integral controller for a minimum phase high order system by using the Lambert W function. *2017 IEEE 15th International Symposium on Intelligent Systems and Informatics (SISY)*, Serbia. IEEE, 2017.
12. Chyi H., Yi. C.C., A numerical algorithm for stability testing of fractional delay systems, *Automatica*, 42, 825–831, 2006.

13. Shinozaki H., Mori T., Robust stability analysis of linear time-delay systems by Lambert W function: Some extreme point results, *Automatica*, 42, 1791–1799, 2006.
14. Yi S., Ulsoy A. G., Nelson P.W., Proportional-integral control of first-order time-delay systems via eigenvalue assignment, Preprint Submitted to IEEE Transactions on Control Systems Technology.

10 Neural Networks– Based Transmission Line Congestion Analysis of Electric Power Systems

Pradyumna Kumar Sahoo
GIET

Prateek Kumar Sahoo
S'O'A University

Prasanta Kumar Satpathy
CET

CONTENTS

10.1 INTRODUCTION

Electrical power systems usually operate in a stressful condition due to several uncertain conditions such as uncertainty in the loading pattern and occurrence of contingencies or faults. Thus, the transmission lines moreover remain congested so far as power transmission is concerned. Congestion level in transmission lines may be explained as a factor or index that describes the relative order of existing loading level as compared with the maximum power transfer capacity. Reduction of the gap between these two power levels leads to increase in congestion. The reasons behind congestion could be (1) over-utilization of generation/transmission capacities, (2) unprecedented abnormal hike in the demand, and (3) unforeseen contingency conditions that might occur anytime, anywhere in the system. As a result, congestion may lead to system collapse due to the cascading events and tripping of critical lines due to overstressing.

The issues related to congestion of transmission lines in electrical power systems are addressed in the literature. Canizares et al. [1] tried to find the effect of congestion management in electricity market pricing. Antonio et al. [2] evaluated the impact of congestion for the purpose of ensuring voltage stability. Yet, the problem poses serious concern for every active researcher in this area. The major trouble that surfaces in the process of evaluation of the transmission line congestion in complex systems/critical conditions is that the conventional load flow technique often suffers from either slower convergence or nonconvergence. Also, it requires repeat execution of the same for each scenario/operating condition.

The major objective of this chapter is to overcome these limitations of conventional load flow approach by the application of neural networks. As observed in the literature [3–5], neural networks show consistently good results in various types of studies concerning to power system analysis such as power system restoration and voltage stability monitoring. During training, the neural network learns from the events, by way of experiencing the training process so that the trained network can be exposed to any set of input data in a future time for necessary validation and testing. Hence, a trained neural network reduces the operator's burden of conducting repeat execution of conventional load flow programs.

10.2   LINE CONGESTION INDEX

The training of neural networks essentially requires an extensive set of input data and a specific target dataset. The input dataset in this chapter covers the line congestion index (LCI) values of all the lines in the test system corresponding to a diverse set of operating conditions. In order to meet this requirement, the authors performed Newton–Raphson load flow simulation, which is used once in the beginning for evaluating the LCI and formation of the input dataset. The procedure adopted in this chapter for evaluation of LCI is enumerated below. The differential voltage and current expressions over an elementary section dx at a distance x from the receiving end of a long transmission line with distributed parameters are represented in Equations (10.1) and (10.2), where z and y imply the impedance and admittance values per unit length of the line, respectively.

$$dV_x = (I_x z) x \qquad (10.1)$$

$$dI_x = (V_x y) x \qquad (10.2)$$

The solution of these expressions results in Equations (10.3) and (10.4):

$$V_x = \left(\frac{V_r + \left(\sqrt{z/y} \right) I_r}{2} \right) e^{\left(\sqrt{z/y} \right) x} + \left(\frac{V_r - \left(\sqrt{z/y} \right) I_r}{2} \right) e^{\left(-\sqrt{z/y} \right) x} \qquad (10.3)$$

$$I = \left(\frac{V_r / \left(\sqrt{z/y} \right) + I_r}{2} \right) e^{\left(\sqrt{z/y} \right) x} + \left(\frac{V_r / \left(\sqrt{z/y} \right) - I_r}{2} \right) e^{\left(-\sqrt{z/y} \right) x} \qquad (10.4)$$

The term $\sqrt{z/y}$ in the above equations represents the characteristic impedance or surge impedance (Z_c) of the line. In case of a particular power system as shown

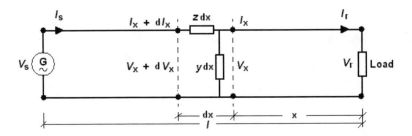

FIGURE 10.1 Generalized representation of a long transmission line (distributed parameters).

in Figure 10.1, while the load impedance matches exactly with the characteristic impedance or surge impedance of the line, the line gets eventually terminated by its own characteristic impedance.

During this condition, maximum power transfer to the load takes place that eventually matches with the limiting power transfer capability of the said line. This limiting power margin is called surge impedance loading (SIL) of the line, as expressed in Equation (10.5):

$$SIL = \frac{|V_l|^2}{|Z_c|} \tag{10.5}$$

The line congestion in transmission systems has been addressed by various researchers. Aswani et al. [6] defined the transmission congestion distribution factor as a ratio of two powers at two candidate lines subject to unit change of power at one of them. However, the approach is limited to show the congestion level of the line subject to maximum change power admissible for the other line. Yingzhong et al. [7] proposed a sensitivity index for wind curtailment, subject to transmission congestion as the ratio of change in curtailed wind generation to that of transmission line capacity. This index, though serves the purpose of deciding wind power generation curtailment, suffers from determining the congestion level of individual lines from a loading perspective. The LCI proposed by Panda et al. [8] is used in this chapter that takes care of both the aforementioned limitations. The proposed LCI is defined as the ratio of the actual value of real power (P_l) transmitted in a particular line to its own SIL, which serves as an indicator of the congestion level in that particular line, as shown in Equation (10.6):

$$LCI = \frac{P_L}{SIL} \tag{10.6}$$

10.3 CONGESTION ANALYSIS BY FEED FORWARD NEURAL NETWORKS

Neural networks offer several applications in solving complex real-world problems with ease. Such applications have been widely accepted by the researchers in the area of electrical power systems. Pandey et al. [9] performed the congestion

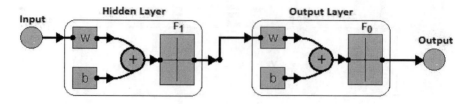

FIGURE 10.2   Generalized structure of FFNN with one hidden layer.

management study using radial basis function neural network (RBFNN) for price prediction of electricity. Alberto et al. [10] demonstrated the possibility of applying some elementary neural network techniques for congestion analysis of a zonal market. Application of feed forward neural networks (FFNN) to power systems has also been reported by other researchers [11–13].

A generalized structure of FFNN with hidden layers is presented in Figure 10.2. This chapter explores the possibility of applying neural network technique for monitoring the congestion level in electrical power transmission networks by considering FFNN. In case of FFNN, there could be more than one hidden layer, and the function blocks (F_1 and F_0) could assume either linear or nonlinear functions.

The hidden layers contain neuron-like elements having interconnectivity. These interconnections by and large determine the network transfer function. Each connection is associated with an index called weight parameter that modulates/transforms the input in accordance with the weighting index. The transfer functions for each hidden layer may differ from one another. In Figure 10.2, the relationship between the output (a) and input (p) for a particular layer is expressed by the transfer function (f) taking into consideration the weight (w) and bias (b) associated to the neurons in that layer, as shown in Equation (10.7). Then the error (e) is calculated as the difference between the specified target vector (T) and output vector (a) as shown in Equation (10.8). The procedure for selection of target vector for this study is described in the next section.

$$a = f\left(wp+b\right) \qquad (10.7)$$

$$e = T - a \qquad (10.8)$$

10.4   IMPLEMENTATION AND RESULT ANALYSIS

The basic objective behind this work is to train the neural networks by way of exposing them to a set of input data reflecting the trend and variations in congestion level of a system pertaining to various practical situations as depicted in Table 10.1. In order to accomplish this goal, the authors have formulated an extensive set of input data containing LCI results obtained from the conventional Newton–Raphson load flow for various operating scenario/events. These operating conditions have been simulated considering various loading levels described as a function of load increasing parameter (λ) and contingency constraints. In the base case loading condition,

TABLE 10.1

Operating Scenarios for Evaluation of LCI

Nomenclature of Scenario/Event	Loading Parameter ($\lambda = 0$ Implies Base Case)	Contingency Constraints
1 to 16	$\lambda = 0$ to 1.5 (0.1 rise per step)	No line outage
17 to 32	$\lambda = 0$ to 1.5 (0.1 rise per step)	Line outage considered (15% of total lines out)

TABLE 10.2

Formation of Study Cases for Performance Comparison

	Type of Neural	Transfer Function Assigned to Layers	
Study Case	Network	Hidden Layer (F_1)	Output Layer (F_0)
Case 1	FFNN	Purely linear	Tan-sigmoid
Case 2	FFNN	Purely linear	Log-sigmoid
Case 3	FFNN	Tan-sigmoid	Purely linear
Case 4	FFNN	Tan-sigmoid	Log-sigmoid
Case 5	FFNN	Log-sigmoid	Purely linear
Case 6	FFNN	Log-sigmoid	Tan-sigmoid

the load increasing parameter takes a zero value, whereas higher values refer to increased loading condition.

In order to simulate the load growth in the system, the complex load demand at the buses has been increased simultaneously as a function of the base loading. This is performed with the help of a load increasing parameter (λ) that gives the loading pattern $S = (1+\lambda)(S_{base})$. Further, the line outage contingency is also considered by imposing a maximum of 15% of existing lines taken for outage. Since the IEEE 30-bus system contains 41 lines, this chapter considers a maximum of 6 line outage contingencies (Table 10.2).

In view of Table 10.1, various situations have been framed for formation of input dataset and the target dataset in order to perform the proposed training analysis. With these considerations, the proposed methodology has been tested on the standard IEEE 30-bus test system. In this chapter, the analysis has been performed to impart successful training to the neural networks through MATLAB simulation. The convergence results thus obtained are presented in Table 10.3, and the plots are shown in Figure 10.3. The overall advantage of applying neural networks for monitoring congestion level is also observable from the contrasting results of conventional approach that shows a significantly high value of convergence time (in minutes).

In view of the above analysis, it is inferred that a suitably trained neural network is capable of monitoring line congestion level in electrical power systems.

TABLE 10.3
Training Results Showing Successful Training Convergence

		Results for Successful Training Convergence		
Study Case	Number of Training Cycles	Number of Epochs	Time Taken (s)	Relative Speed of Convergence as Compared with Case 12
Case 1	118	1172	49.07	8.47
Case 2	36	360	19.90	20.94
Case 3	10	98	5.39	77.18
Case 4	66	660	32.26	12.89
Case 5	93	929	41.32	10.06
Case 6	27	270	13.35	31.16

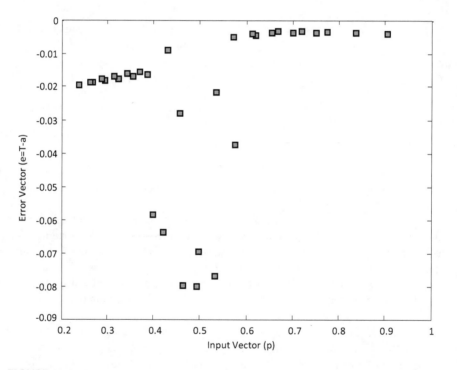

FIGURE 10.3   Error–input distribution obtained from a particular training convergence of FFNN.

10.5   CONCLUSIONS

The basic objective of this chapter is satisfied through the case study conducted on IEEE 30-bus test system. The study is based on a conventional approach of load flow simulation for formulation of the input dataset consisting of LCI values for 32 situations/events. The dataset so formed are presented to three types of neural networks for imparting successful training. The study validates that all the three types of neural networks (FFNN, ENN, RBFNN) used in the study are capable of acquiring successful training showing different training performance. The training performance comparison indicates that RBFNN scheme is the most compact and robust network among the three networks that offer fastest convergence.

REFERENCES

1. Canizares, C.A., Hong, C., Milano, F., Singh, A.: Transmission congestion management and pricing in simple auction Electricity markets, *International Journal Emerging Electric Power Systems*, Vol. 1(1) (2004) 1–10.
2. Antonio, J., Conejo, Milano, F., Raquel, G.: Congestion management ensuring voltage stability, *IEEE Transaction on Power Systems*, Vol. 21(1) (2006) 357–364.
3. Iman, S., Abbas, K., Rene, F.: Radial basis function neural network appln to power system restoration studies, *Computational Intelligence and Neuroscience*, Vol. 1 (2012) 1–10.
4. Bahamanyar, A.R., Karami, A.: Power system voltage stability monitoring using artificial neural networks with a reduced set of inputs, *Electrical Power and Energy Systems*, Vol. 58. (2014) 246–256.
5. Zhou, D.Q., Annakkage, U.D., Rajapakse, A.D.: Online monitoring of voltage stability margin using an artificial neural network, *IEEE Transaction on Power Systems*, Vol. 25(3) (2010) 1566–1574.
6. Aswani, K., Srivastava, S.C., Singh, S.N.: A zonal congestion management approach using real and reactive power rescheduling, *IEEE Transaction on Power Systems*, Vol. 19(1) (2004) 554–562.
7. Yingzhong, G., Xie, L., Brett, R., Hesselbaek, B.: Congestion-induced wind curtailment: Sensitivity analysis case studies, *Proc. of North American Power Symposium* (2011) 1–7, USA.
8. Panda, R.P., Sahoo, P.K., Satpathy, P.K., Paul, S.: Analysis of critical conditions in electric power systems by feed forward and layer recurrent neural networks, *International Journal of Electrical Engineering and Informatics*, Vol.6(3) (2014) 447–459.
9. Pandey, S.N., Tapaswi, S., Srivastava, L.: Price prediction based congestion management using growing RBF neural network, *Proc. Annual IEEE India Conference* (2008) 482–487, New Delhi.
10. Alberto, B., Maurizio, D., Marco, M., Marco, S.P., Politecnico, M.: Congestion management in a zonal market by a neural network approach, *European Transaction on Electrical Power*, Vol. 19(4) (2009) 569–584.
11. Xue, L., Jia, C., Dajun, D.: Comparison of Levenberg-Marquardt method and Path Following interior point method for the solution of optimal power flow problem, *Emerging Electric Power Systems*, Vol. 13(3) (2012) 15–35.
12. Ilamathi, B., Selladurai, V.G., Balamurugan, K.: ANN-SQP approach for NOx emission reduction in coal fired boilers, *Emerging Electric Power Systems*, Vol. 13(3) (2012) 1–14.
13. Keib, A.A.E., Ma, X.: Application of artificial neural networks in voltage stability assessment, *IEEE Transaction on Power Systems*, Vol. 10(4) (1995) 1890–1896.

11 Study of Dielectric and Conductivity Behavior of PMMA–rGO Polymer Nanocomposites

T. Badapanda
CV Raman College of Engineering

S.R. Mishra
Gandhi Institute for Education and Technology

S. Parida
CV Raman College of Engineering

CONTENTS

11.1 INTRODUCTION

High-permittivity polymer-based composites are highly desired due to their potential applications as high-energy density capacitors, electroluminescent devices, gate dielectrics, electrical energy storage, and inherent advantages of easy processing, low temperature processing conditions, flexibility, and lightweight [1,2]. However, these polymer films generally have low dielectric constants, and thus yield low performance in electronic devices. One strategy to increase the dielectric constant of polymer dielectric films is to use ferroelectric ceramic (e.g., barium titanate or strontium titanate) and/or electrically conductive fillers (e.g., metal nanoparticles or carbon-based nanostructures). Incorporating these fillers in the polymer matrix can increase the capacitance due to the induced interfacial or space charge polarization between the filler and the polymer which opened a new dimension for the production of lightweight, low-cost, and high-performance composite materials for a range of applications [3–6].

Among polymers, poly(methyl methacrylate) (PMMA) is an amorphous polymer which does not crystallize due to its semi-flexible and rigid backbone structure. The high stiffness of PMMA along with its biocompatibility is commonly used in materials for bone substitute applications. Despite its exceptional mechanical behavior, PMMA exhibits a rather low dielectric permittivity. PMMA has been the subject of numerous composite studies focusing on the improvement of strength and durability. Although polymers possess relatively low permittivity, they can withstand high fields due to their high dielectric strength. In order to enhance the permittivity of the polymers, particulate composites possessing high permittivity have become increasingly important for capacitor applications [7–12]. Compared with carbon nanotubes (CNTs), graphene has a higher surface-to-volume ratio because of the inaccessibility of the CNT's inner tube surface to polymer molecules. This makes graphene potentially more favorable for improving the properties of polymer matrices, such as electrical properties [13]. One of the advantages of graphene oxide (GO) or reduced grapheme over graphene is that it can be easily dispersed in water and physiological environments due to its abundant hydrophilic groups, which include hydroxyl, epoxide, and carboxylic groups on its large surface [14,15]. Good dispersion is crucial for achieving the desired enhancement in the final physical and chemical properties of the composites [16], especially for graphene, which has a strong tendency to agglomerate due to intrinsic van der Waals forces [17]. Reduced graphene oxide (rGO) shows enhanced dielectric constant with low dielectric loss as compared with CNT due to the increase of interfacial polarization at the interface between the polymer and filler [18]. Several studies have reported the successful incorporation of graphene nanosheets into the PMMA matrix with different preparation techniques using various methods of graphene preparation [19–22].

Several articles have demonstrated the successful incorporation of graphene fillers into the PMMA matrix. The results show that at only 1 wt% of graphene loading, remarkable changes are obtained in the elastic modulus and tensile strength and electrical conductivity [23–25]. Many factors including filler aspect ratio, surface area, concentration, dispersion state, mode of synthesis, and contact resistance are key factors affecting the electrical properties of graphene/polymer composites [26,27]. It is reported that the electrical conductivity and the specific surface area of fillers are strongly related to the dielectric properties of composite films containing conductive fillers. High-conductivity fillers increase interfacial polarization at the interface between the polymer and the filler [28]. In addition, increased specific surface area of fillers increases the density of the interfacial polarization [29]. Increasing both parameters, which contribute to greater polarization, can improve the dielectric constant and conductivity of the polymer composite film. From these factors, we expect that the incorporation of rGO in the PMMA matrix will enhance the dielectric and conductivity properties of PMMA.

11.2   MATERIAL SYNTHESIS

11.2.1   SYNTHESIS OF REDUCED GRAPHENE OXIDE (rGO)

The modified Hummers method was used for the synthesis of GO from graphite powder (SDFCL, 100 μm) as the starting precursor. Initially, graphite powder was mixed with a strong oxidizing agent using concentrated sulfuric acid and potassium permanganate. The temperature of the suspension was maintained at 20°C and stirred for 2 h. Then deionized water (Millipore) and 30% hydrogen peroxide were mixed in the solution to terminate the above reaction. Washing of the suspension with 1:10 HCl solution was undertaken to remove Mn-based oxides and metal ions and then filtered. The collected paste from the filter paper was dried at 50°C–60°C until it became agglomeration. The agglomeration was dispersed into deionized water using ultrasonication. GO contains a range of reactive oxygen functional groups, which render it a good candidate for use in the electronic device applications. Therefore, it is very much essential to reduce the highly resistive GO, and it was chemically reduced by sodium borohydride ($NaBH_4$), a strong reducing agent. In a typical synthesis, 0.7566 g of GO was dispersed in 20 ml of 1 M freshly prepared aqueous $NaBH_4$ solution under vigorous stirring for 10 min. Then, the rGO powder was collected by filtration.

11.2.2   PREPARATION OF PMMA–rGO COMPOSITE

PMMA with a repeated molecular weight of 100.12 g/mole and toluene were purchased from L.G. Chem Ltd. and Merck India and were used as the precursor and its solvent, respectively. The PMMA solution was prepared with the concentration of 1 g/10 ml at 40°C and is maintained throughout all composites. In this method, initially rGO with different wt% were dispersed in PMMA solution and ultrasonicated for 30 min. Final solution was stirred vigorously for 12 h at 40°C until a homogenous solution was obtained. The composite solution was poured on hard aluminum foil and left to dry at room temperature. The composite's thick films thus were obtained. The photograph of the PMMA–rGO composite film is shown in Figure 11.1.

11.3   CHARACTERIZATIONS

X-ray diffraction (XRD) patterns were recorded on a diffractometer (Rigaku, DMax 2000PC, Japan) using Cu-Ka radiation. FT-Raman spectroscopy measurements of GO and rGO were performed using an RFS100/S FT-Raman spectrometer (Bruker, Germany). The Fourier-transformed infrared (FT-IR) absorption spectrum was recorded using FT-IR spectrometer (IR-Prestige 21, SHIMADZU, Japan). The microstructure of ceramics, sintered at 1,150 1C for 4 h, was observed by scanning electron microscopy (SEM) (JEOL:JSM 6480LV, Japan). The frequency - (1 Hz–1 MHz) and

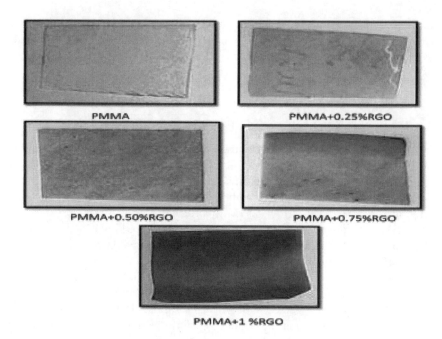

FIGURE 11.1   Photograph of PMMA–rGO composites.

temperature-dependent dielectric and impedance measurements were carried out using a N4L-NumetriQ (model PSM1735) connected to a computer.

11.4   RESULT AND DISCUSSION

11.4.1   XRD, FTIR, AND RAMAN SPECTROSCOPY OF rGO

Figure 11.2a shows the XRD pattern of rGO. A typical broad peak near at 26° is observed which corresponds to (002) orientation. This broad peak indicates the removal of oxygen functional groups associated with GO. It also suggests that the chemical oxidation of graphite was successfully carried out followed by reduction to obtain graphene (rGO). Figure 11.2b shows the FTIR spectra of GO and rGO. The FTIR spectra of GO represents the characteristic peaks corresponding to the oxygen functionalities, that is, C–O–C, C–OH, and C=O at 1,225, 1,030, and 1,718 cm^{-1}, respectively, and the peaks at 3,440 and 1,396 cm^{-1} correspond to hydroxyl (–OH) groups and the broad range from 2,500 to 3,800 cm^{-1} confirms the presence of carboxylic (–COOH) group [30].

For rGO, Figure 11.3 indicates that the O–H band at 3,410 cm^{-1} was reduced in intensity due to the deoxygenation of the GO-oxygenated functionalities. The spectrum of rGO also contains bands at 1,620 and 1,225 cm^{-1}, which correspond to C=C and C–O groups, respectively. The high intensity of the main peaks in GO confirms the presence of a large amount of oxygen functional groups after the oxidation process. After the GO was reduced, hydroxyl and alkoxy groups were significantly

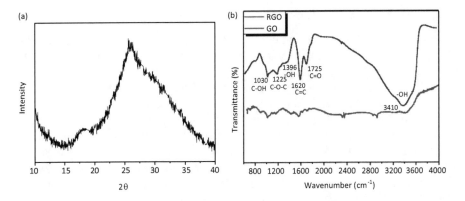

FIGURE 11.2   (a) X-ray diffraction of rGO. (b) FTIR spectra of GO and rGO.

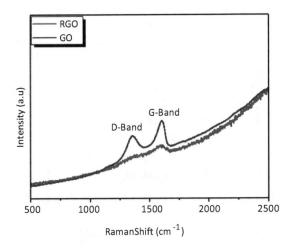

FIGURE 11.3   Raman spectra of GO and rGO.

decreased, and the phenol C=C ring stretching at 1,585 cm^{-1} was present confirming the formation of rGO [31,32].

Raman spectroscopy was also employed to characterize the rGO. Raman spectroscopy is a very powerful tool that provides essential information for evaluating the covalent modification of rGO nanosheets [33].

The Raman spectrum for rGO and GO displays characteristic _D' and _G' bands at 1,350 and 1,601 cm^{-1}, respectively (Figure 11.2b). The D-band for GO is more intense than G-band (ID/IG ~ 0.80), which shows there are more sp^3 hybridized carbons formed due to generation of –OH, –COOH, and epoxide groups during oxidation than sp^2 hybridized carbons. The D-band is related to the sp^3 state of carbon which gives a proof of disruption of the aromatic p-electrons (sp^2 hybridized) of rGO sheet. For rGO, the intensity ratio of these two bands (i.e., ID/IG ratio) is found to be 0.76 [34].

11.4.2   XRD AND FTIR OF COMPOSITE

From Figure 11.4, it is observed that pure PMMA possesses broad peaks at $2\theta = 15°$ and 35°, whereas with incorporation of rGO-loaded PMMA composite, only the diffraction peak of PMMA is observed and rGO hump is suppressed. The above results demonstrate that rGO sheets are fully exfoliated and dispersed in the polymer matrix. Furthermore, FTIR and XRD were employed to characterized PMMA–rGO nanocomposite. Figure 11.5 shows PMMA and PMMA–rGO composite. It can be seen that there is a distinct absorption band from 1,151 to 1,244 cm^{-1}, which can be attributed to the C–O–C stretching vibration. The two bands at 1,388 and 751 cm^{-1} can be attributed to the a-methyl group vibrations. The band at 987 cm is the characteristic absorption vibration of PMMA, together with the bands at 1,062 and 843 cm^{-1}. The band at 1,725 cm^{-1} shows the presence of the acrylate carboxyl group. The band at 1,444 cm^{-1} can be attributed to the bending vibration of the C–H bonds of the –CH$_3$ group. The bands at 2,950 cm^{-1} can be assigned to the C–H bond stretching vibrations of the –CH$_2$– groups. Furthermore, there is a weak absorption band at 1,641 cm^{-1}, which can be attributed to the –OH group bending vibrations of physisorbed moisture. On the basis of the above discussions, it can be concluded that the prepared polymer was indeed macromolecular PMMA [35].

The broad humps centered at 3,410 cm^{-1} correspond to the O–H group and are observed from RGO-0.25 to RGO-1.0 nanocomposites (see square bracket in Figure 11.3). It indicates rGO is well dispersed in PMMA matrix. Since the relative intensity of rGO is very less as compared with PMMA, some of the rGO peaks are merged in peaks of PMMA matrix [36].

11.5   DIELECTRIC PROPERTIES OF RGO–PMMA COMPOSITES

Figure 11.6 shows the dielectric permittivity (ε') of the rGO–PMMA composites as a function of frequency at room temperature. It is observed from the figure that the dielectric constant decreases with increase in frequency and becomes constant

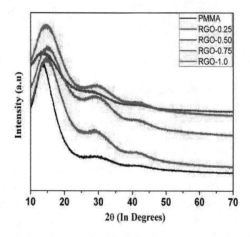

FIGURE 11.4 X-ray diffraction pattern of PMMA–rGO composite.

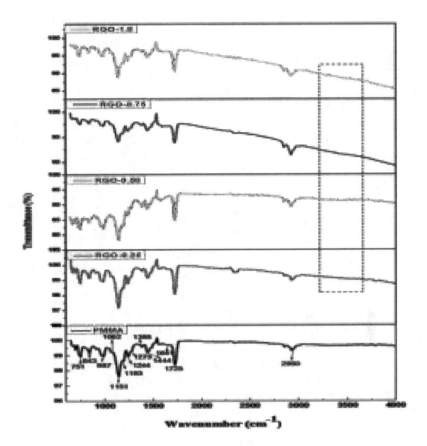

FIGURE 11.5 FTIR spectra of PMMA–rGO composite.

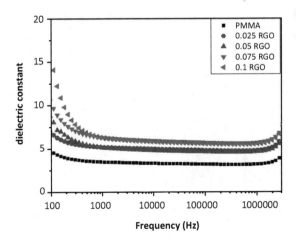

FIGURE 11.6 Frequency-dependent dielectric behavior of PMMA–rGO composite at room temperature.

at high frequency. In addition, the fact that for polar materials the initial value of dielectric permittivity is high, but as the frequency of the field is raised, the value begins to drop due to the disability of the dipoles to follow the field variations at high frequencies and also due to electrode polarization effects. At high frequencies, the periodic reversal of the electric field occurs so fast that there is no excess ion diffusion in the direction of the field. The polarization due to the charge accumulation decreases, leading to the decrease in the value of dielectric permittivity [37]. As expected, dielectric properties of the rGO–PMMA composites show an enhanced volume fraction of rGO incorporated within the PMMA matrix. It is well-documented that the increment in dielectric permittivity of conductive filler/polymer is attributed to the so-called microcapacitor effect [38–40]. In the present work, the conductive rGO isolated by very thin insulating polymer layers form lots of microcapacitor structures throughout the composites, which can increase in the intensity of local electric field [41]. The increase in the intensity of local electric field can further promote the migration and accumulation of charge carriers at the interfaces between the rGO and the PMMA matrix. This interfacial polarization, also known as the Maxwelle–Wagnere–Sillars polarization, is the major cause for the enhancement of dielectric permittivity of the rGO–PMMA composites.

Figure 11.7 shows the variation dielectric loss of rGO–PMMA composite with frequency at room temperature. It is observed that the dielectric loss increases with increase in rGO content. Such increase in the loss factor is the inevitable consequence of the significantly enhanced conductivity of the rGO–PMMA composites and could be considered as one important feature of the percolative composites. For composites incorporated with conductive fillers, the dielectric loss is mainly caused by the leakage current in the composites. Higher content of conductive fillers could construct more conductive pathways, and thus result in more significant leakage current and dielectric loss. The main contribution to the conductive processes could be

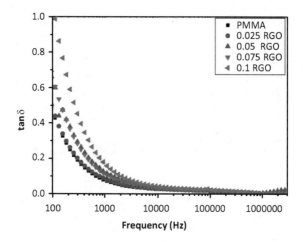

FIGURE 11.7 Frequency-dependent dielectric loss of PMMA–rGO composite at room temperature.

attributed to space charge in the sample and probably arises from trapped charges in the rGO–PMMA interfaces.

11.5.1 IMPEDANCE SPECTROSCOPY STUDY

Figure 11.8 shows the variation of real part of impedance (Z') of rGO–PMMA as a function of frequency at room temperature. The nature of variation shows a monotonous decrease in the value of with rise in the frequency and becomes constant at high frequency. The decrease in impedance with a rise in frequency also indicates a possibility of increase in the AC conductivity with increases in temperature and frequency. The merger of real part of impedance (Z') in the high-frequency domain for all temperatures indicates a possibility of the release of space charge as a result of lowering in the barrier properties of the material. It is also observed that the value increases as the rGO content increases.

Figure 11.9 shows the variation of the imaginary part of the impedance with frequency at room temperatures for rGO–PMMA composites. The frequency patterns exhibit some important features such as decrease in the magnitude of Z'' toward higher frequency. The effect of increase in ceramic substitution on the electrical behavior of the samples can clearly be seen in terms of (1) an increase in the magnitude of Z''. As the concentration of rGO increases, there is an increase in value. This result also shows a significant effect of rGO substitution on the electrical behavior of composite.

11.5.2 AC CONDUCTIVITY ANALYSIS

Figure 11.10 shows the variation of AC electrical conductivity (ζ_{ac}) of PMMA–rGO composites as a function of frequency at room temperatures. The AC conductivity is calculated from the dielectric data using the relation $\zeta_{ac} = \varepsilon_o \varepsilon' \omega \tan\delta$. The low value of AC conductivity in low-frequency region may be due to the less availability of mobile charge carriers for hopping from one localized state to another due

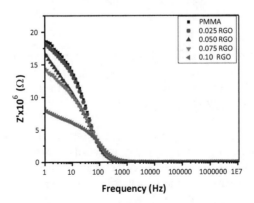

FIGURE 11.8 Frequency-dependent real part of impedance of PMMA–rGO composite at room temperature.

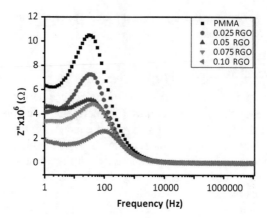

FIGURE 11.9 Frequency-dependent imaginary part of impedance of PMMA–rGO composite at room temperature.

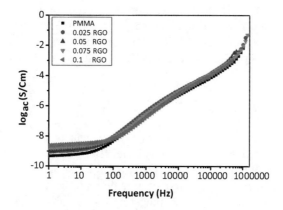

FIGURE 11.10 Frequency-dependent AC conductivity of PMMA–rGO composite at room temperature.

to pronounced polarization effect occurring at the electrode and interfaces in this frequency region. With increase in frequency, the nonlinearity in these curves has been observed with increased value of slope depicting the increase in σ_{ac}. The rise in values of AC conductivity with frequency may be attributed to increase in mobility of charge carriers.

A convenient formalism to investigate the frequency behavior of conductivity at constant temperature in a variety of materials is based on the power relation proposed by Jonscher (), (3) where $\zeta(\omega)$ is the total conductivity, ζ_{dc} is the frequency-independent conductivity, and the temperature-dependent pre-exponential factor A and frequency exponent n ($0 \leq n \leq 1$) are the temperature and material intrinsic property–dependent constants. The term $A\omega^n$ is frequency dependent and characterizes all the dispersion phenomena. The conductivity spectrum displays characteristic conductivity dispersion throughout the frequency range. The frequency-independent plateau at a low frequency for higher temperatures is attributed to the long-range translational motion of ions contributing to DC conductivity. The conductivity increases with increase in the rGO content.

11.6   CONCLUSION

In summary, rGO–PMMA composite was synthesized by solvent casting technique. The rGO was prepared by the modified Hummer method. The XRD, FTIR, and Raman spectroscopy analyses confirm the preparation of rGO. The XRD and FTIR studies of the composite show that the rGO is incorporated in the polymer matrix. The micrographs of polymer composites show well dispersion of rGO in the polymer matrix. The dielectric properties of the polymer composites were studied with variation of frequency at room temperature. It was observed that the dielectric constant increases with increase in rGO content. The enhanced dielectric properties of the rGO composites were attributed to the induced polarization effect and intrinsic high dielectric constant of rGO. The impedance spectroscopy analysis was carried out to study the conduction mechanism and it is obtained that the real and imaginary impedance value increases with rGO content. The AC conductivity with variation in frequency was studied at room temperature and it is found that the AC conductivity increases with rGO content due to the tunneling effect. The enhanced dielectric and conduction properties of the polymer composite makes the material suitable for industrial application.

ACKNOWLEDGEMENT

T. Badapanda acknowledges the financial support from the Council of Scientific and Industrial Research, grant number 80(0084)/14/EMR-II.

REFERENCES

1. J.-Y. Kim, T. Y. Kim, J. W. Suk, H. Chou, J.-H. Jang, J. H. Lee, I. N. Kholmanov, D. Akinwande, and R. S. Ruoff, Enhanced dielectric performance in polymer composite films with carbon nanotube-reduced graphene oxide hybrid filler. *Small*, 10(16), 3405–3411 (2014).
2. J. Y. Kim, S. H. Park, T. Jeong, M. J. Bae, Y. C. Kim, I. Han, and S. Yu, High electroluminescence of the ZnS:Mn nanoparticle/cyanoethyl-resin polymer/single-walled carbon nanotube composite using the tandem structure. *J. Mater. Chem.*, 22(38), 20158–20162 (2012); S. Y. Yang, K. Shin, C. E. Park, *Adv. Funct. Mater.* 2005, 15(11), 1806–1814.
3. P. Noorunnisa Khanam, Deepalekshmi Ponnamma, and M. A. AL-Madeed, Electrical properties of graphene polymer nanocomposites. K. K. Sadasivuni et al. (eds.), *Graphene-Based Polymer Nanocomposites in Electronics*, 25–47. Springer, Cham (2015). doi:10.1007/978-3-319-13875-6_2.
4. T. T. Zhang, J. H. Yang, N. Zhang, T. Huang, and Y. Wang, Achieving large dielectric property improvement in poly(ethylene vinyl acetate)/thermoplastic polyurethane/multiwall carbon nanotube nanocomposites by tailoring phase morphology. *Ind. Eng. Chem. Res.*, 56(13), 3607–3617 (2017).

5. H. J. Mao, T. T. Zhang, T. Huang, N. Zhang, Y. Wang, and J. H. Yang, Fabrication of high-k poly(vinylidene fluoride)/ Nylon 6/carbon nanotube nanocomposites through selective localization of carbon nanotubes in blends. *Polym. Int.*, 66(4), 604–611 (2017).

6. G. S. Kumar, D. Vishnupriya, K. S. Chary, and T. U. Patro, High dielectric permittivity and improved mechanical and thermal properties of poly (vinylidene fluoride) composites with low carbon nanotube content: effect of composite processing on phase behavior and dielectric properties. *Nanotechnology*, 27(38), 385702 (2016).

7. K. Goto, M. Hashimoto, H. Takadama, J. Tamura, S. Fujibayashi, and K. Kawanabe, Mechanical setting and biological properties of bone cements containing micron-sized titania particles. *J. Mater. Sci. Mater. Med.*, 19(3), 1009 (2008).

8. A. Boger, A. Bisig, M. Bohner, P. Heini, and E. Schneider, Variation of the mechanical properties of PMMA to suit osteoporotic cancellous bone. *J. Biomater. Sci. Polym. Ed.*, 19(9), 1125 (2008).

9. J. N. Coleman, U. Khan, and Y. K. Gun'ko, Mechanical reinforcement of polymers using carbon nanotubes. *Adv. Mater.*, 18, 689 (2006).

10. Z. M. Li, et al., A novel approach to preparing carbon nano tube reinforced thermoplastic polymer composites. *Carbon*, 43, 2397 (2005).

11. B. Marrs, R. Andrews, T. Rantell, and D. Pienkowski, Augmentation of acrylic bone cement with multiwall carbon nanotubes. *J. Biomed. Mater. Res.*, 77A, 269 (2006).

12. Y. Ul-Haq, I. Murtaza, S. Mazhar, R. Ullah, M. Iqbal, Zeeshan-ul-Huq, A. A. Qarni, and S. Amin, Dielectric, thermal and mechanical properties of hybrid PMMA/RGO/Fe2O3 nanocomposites fabricated by in- situ polymerization. *Ceram. Int.*, 46, 5828–5840 (2020). doi:10.1016/j.ceramint.2019.11.033.

13. J. Du, and H. M. Cheng, The fabrication, properties, and uses of graphene/polymer composites. *Macromol. Chem. Phys.*, 213, 1060–1077 (2012).

14. S. Pasupuleti, and G. Medras, Ultrasonic degradation of poly (styrene-co-alkyl methacrylate) copolymers. *Ultrason. Sonochem.*, 17, 819–826 (2010).

15. C. Mehmet, and P. Seven, Synthesis, Characterization and investigation of dielectric properties of two- armed graft copolymers prepared with methyl methacrylate and styrene onto PVC using atom transfer radical polymerization. *Reac. Funct. Polym.*, 71, 395–401 (2011).

16. D. R. Paul, and L. M. Robeson, Polymer nanotechnology: Nanocomposites. *Polymer*, 49, 3187–3204 (2008).

17. J. R. Potts, D. R. Dreyer, C. W. Bielawski, and R. S. Ruoff, Graphene-based polymer nanocomposites. *Polymer*, 52, 5–25 (2011).

18. Z. M. Dang, Y. H. Zhang, and S. C. Tjong, Enhanced dielectric performance in polymer composite films with carbon nanotube-reduced graphene oxide hybrid filler. *Synthetic Met.*, 146(1), 79–84 (2004).

19. J. Y. Jang, M. S. Kim, H. M. Jeong, and C. M. Shin, Graphite oxide/poly(methyl methacrylate) nanocomposites prepared by a novel method utilizing macroazoinitiator. *Comp. Sci. Tech.*, 69, 186–191 (2009).

20. X. Y. Yuan, L. L. Zou, C. C. Liao, and J. W. Dai, Improved properties of chemically modified graphene/poly(methyl methacrylate) nanocomposites via a facile in-situ bulk polymerization. *Express Polym. Lett.*, 6, 847–858 (2012).

21. J. Wang, H. Hu, X. Wang, C. Xu, M. Zhang, and X. Shang, Preparation and mechanical and electrical properties of grapheme nanosheets–poly(methyl methacrylate) nanocomposites via in situ suspension polymerization. *J. App. Poly. Sci.*, 122, 1866–1871 (2011).

22. X. Zeng. J. Yang, and W. Yuan, Preparation of a poly(methyl methacrylate)-reduced grapheme oxide composite with enhanced properties by a solution blending method. *Eur. Polym. J.*, 48, 1674–1682 (2012).

23. Z. Fan, F. Gong, S. T. Nguyen, and H. M. Duong, Advanced multifunctional graphene aerogel – Poly (methyl methacrylate) composites: Experiments and modelling. *Carbon*, 81, 396–404 (2015). doi:10.1016/j.carbon.2014.09.072.

24. T. Ramanathan, et al., Functionalized graphene sheets for polymer nanocomposites. *Nat. Nanotechnol.*, 3(6), 327–331 (2008).
25. V. H. Pham, T. T. Dang, S. H. Hur, E. J. Kim, and J. S. Chung. Highly conductive poly (methyl methacrylate) (PMMA)-reduced graphene oxide composite prepared by self-assembly of PMMA latex and graphene oxide through electrostatic interaction. *ACS Appl. Mater. Interfaces*, 4(5), 2630–2636 (2012).
26. T. N. Zhou, X. D. Qi, and Q. Fu, The preparation of the poly(vinyl alcohol)/graphene nanocomposites with low percolation threshold and high electrical conductivity by using the large-area reduced graphene oxide sheets. *Express Polym. Lett.*, 7, 747–755 (2013).
27. J. U. Park, S. W. Nam, M. S. Lee, and C. M. Lieber, Synthesis of monolithic graphene–graphite integrated electronics. *Nat. Mater.* 11, 120–125 (2012).
28. B. H. Wang, G. Z. Liang, Y. C. Jiao, A. J. Gu, L. M. Liu, L. Yuan, and W. Zhang, Two-layer materials of polyethylene and a carbon nanotube/cyanate ester composite with high dielectric constant and extremely low dielectric loss. *Carbon*, 54, 224–233 (2013).
29. L. Valentini, D. Puglia, E. Frulloni, I. Armentano, J. M. Kenny, and S. Santucci, Dielectric behavior of epoxy matrix/single-walled carbon nanotube composites. *Com. Sci. Technol.*, 64, 23 (2004).
30. S. N. Tripathi, P. Saini, D. Gupta, and V. Choudhary, Electrical and mechanical properties of PMMA/reduced grapheme oxide nanocomposites prepared via in situ polymerization, *J. Mater. Sci.*, 48, 6223–6232 (2013).
31. M. El Achaby, F. Z. Arrakhiz, S. Vaudreuil, E. M. Essassi, and A. Qaiss, Piezoelectric β-polymorph formation and properties enhancement in graphene oxide—PVDF nano-composite films. *Appl. Surf. Sci.*, 258(19), 7668–7677 (2012).
32. V. Loryuenyong, K. Totepvimarn, P. Eimburanapravat, W. Boonchompoo, and A. Buasri, Preparation and characterization of reduced graphene oxide sheets via water-based exfoliation and reduction methods. *Advances in Materials Science and Engineering*, 2013, 5 pages (2013).
33. J. Wang, H. Hu, X. Wang, C. Xu, M. Zhang, and X. Shang, Preparation and mechanical and electrical properties of graphene nanosheets–poly(methyl methacrylate) nanocomposites via in situ suspension polymerization. *J. App. Poly. Sci.*, 122, 1866–1871 (2011).
34. S. K. Behura, S. Nayak, I. Mukhopadhyay, and O. Jani, Junction characteristics of chemically-derived graphene/p-Si heterojunction solar cell. *Carbon*, 67, 766–774 (2014).
35. G. Duan, C. Zhang, A. Li, X. Yang, L. Lu, and X. Wang, Preparation and characterization of mesoporous zirconia made by using a poly (methyl methacrylate) template, *Nanoscale Res. Lett.*, 3:118–122 (2008). Z.X. Lin, Analysis and Identification of Infrared Spectrum of the Polymer (Sichuan University Press, Chengdu, 1989).
36. M. A. Aldosari, A. A. Othman, and E. H. Alsharaeh, Synthesis and characterization of the in situ bulk polymerization of PMMA containing graphene sheets using microwave irradiation. *Molecules*, 18, 3152–3167 (2013).
37. V. Raja, A. K. Sharma, and V. V. R. N. Rao, Electrical conductivity and dielectric characteristics of in situ prepared PVA. *Mater. Lett.*, 58, 3242–3247 (2004).
38. Z. M. Dang, Y. H. Lin, and C. W. Nan, Characterization of PZT/PVC composites added with carbon black. *Adv. Mater.*, 15, 1625–1629 (2003).
39. Z. M. Dang, L. Wang, Y. Yin, Q. Zhang, and Q. Q. Lei, Giant Dielectric Permittivities in Functionalized Carbon-Nanotube/ Electroactive-Polymer Nanocomposites, *Adv. Mater.*, 19, 852–857 (2007).
40. Z. M. Dang, J.-K. Yuan, J.-W. Zha, T. Zhou, S.-T. Li, and G.-H. Hu, High performance of P(VDF-HFP)/Ag@TiO$_2$ hybrid films with enhanced dielectric permittivity and low dielectric loss. *Prog. Mater. Sci.*, 57, 660–723 (2012).
41. Y. Li, W. Yang, S. Ding, X.-Z. Fu, R. Sun, W.-H. Liao, and C.-P. Wong, Tuning dielectric properties and energy density of poly (vinylidene fluoride) nanocomposites by quasi core– shell structured BaTiO3@ graphene oxide hybrids. *J. Mater. Sci. Mater. Elect.*, 29, 1082–1092 (2018).

12 Concept of Green Computing Linked to Structural Design and Analysis
A Case Study

Binaya Kumar Panigrahi and
Soumya Ranjan Satapathy
Gandhi Institute for Education and Technology
(Affiliated to Biju Patnaik University of Technology)

CONTENTS

12.1 INTRODUCTION

In the present scenario, in every application, computing facility has become an essential requirement for everybody. It makes life smoother and saves a lot of time and human efforts, but the use of computer also increases power consumption and also generates a greater amount of heat. Greater power consumption and greater heat generation means greater emission of greenhouse gases like carbon dioxide that have various harmful impacts on our environment and natural resources. This is because we are not aware about the harmful impacts of the use of computer on environment. In fact, all the peripheral devices including the data centers and networking devices produce a large amount of carbon dioxide emission. Maximum emission of carbon dioxide is obtained from computer peripherals which are not biodegradable. Also, there are toxic chemicals used in the manufacturing of computers and when disposed informally, it badly impacts on the environment. So, to save our environment and to reduce the harmful impacts of computers we have to become aware about the problems related to them. To decrease these impacts the term green computing comes into existence.

Several reasons for implementation of green computing include computers and electronic devices that consume a lot of electricity that have some harmful impacts on our environment. Also, most of electronic devices generate a lot of heat which cause the emission of carbon dioxide. With the rapid increase of carbon dioxide emission, the rate of global warming has increased through anthropogenic climate change. Moreover, the manufacturing of computer products depends heavily on the use of toxic chemicals for electrical insulation, soldering, and fire protection. All these causes can be reduced using green computing. Considering green design aims at designing energy-efficient data centers. Similarly, green manufacturing aims at manufacturing of electronic components, computers, and other associated subsystems with minimal impact on the environment.

12.2   REVIEW OF LITERATURE

As mentioned in Ref. [1], because of the growth in computing needs, increased energy cost and global warming pose challenges to IT industry. Also, the future of green computing will be based on efficiency, rather than reduction in consumption.

In Ref. [2], the major focus of green IT is based on the organization's self-interest in energy cost reduction at data centers. The result of which is the corresponding reduction in carbon generation. The secondary objective of green IT needs to focus beyond energy use in the data centers, and the focus should be on innovation and improving alignment with overall corporate social responsibility efforts.

Soomro et al. [3] in their work have focused on the basic approach of cloud computing delivering information and communication technology services by improving the utilization of data center resources.

Berl et al. [4] provided an energy-efficient technology along with potential significance of energy savings. They focused on only hardware aspects which can be fully explored with respect to system operation and networking aspects. They also observed that cloud computing results in better resource utilization, which is good for the sustainability movement for green technology.

As cited in Ref. [5], the green computing concept can be easily shared and accessed along with the unused resources on idle computer systems. Leveraging the unused computing power of modern machines to create an environmentally proficient substitute to traditional desktop computing is a cost-effective option. This makes it possible to reduce carbon dioxide emissions to minimum and reduce electronic waste up to 80%.

As cited in Ref. [6], the green computing concentrates mainly on the software development without consuming more time and energy along with the selection and usage of hardware which will not need high support of power resources.

As cited in Ref. [7], the process of pre-fetching and caching of instruction in general helps in saving lots of energy. In pre-fetching and caching the instructions which need to be executed for the next process are fetched from its memory and sent to the cache. Thus, the instruction will be stored temporarily and can be accessed directly from the cache without searching in the memory.

According to Ref. [8], the factors linked to individual system can combine many physical systems into a single integrated system and therefore the original hardware and system can be unplugged resulting in reducing power and cooling consumption. Instead of setting a server and a cooling system for that, it will be better to access a big system server in a virtualized manner. The concept of virtualization is best suitable in the green computing area because it can save power and can also cut costs breaking the link between the applications, application components, system services, and storage systems.

Guenach et al. [9] focused on energy management and implementation scheme to minimize energy consumption in telecommunication networks. They observed that it can be divided as per deployment level, traffic routing level, as well as architectural level.

Curtis-Maury et al. [10] proposed an approach to throttle the concurrency in a system to levels with higher predicted efficiency from both performance and energy standpoints. The predictions are based on machine learning, specifically artificial neural networks. The learning process is where a set of multithreaded scientific applications are executed.

Schmidt et al. [11] in their work analyzed the basic aspects based on the potential throughput, average response time, and backlog length. They observed that there is no existing method to estimate the amount of energy linked to the architectures implemented on a given processor.

Hovland et al. [12] proposed a software architecture for embedded computers of different energy equipment, for example, battery and rectifier. The goals of the new architecture were to reduce scalability as well as ensuring incorporation of new communication techniques.

Shirazi [13] discussed about concurrency to increase the application responsiveness. One high-priority thread can handle user requests, while another thread can run concurrently to perform the necessary operations.

Korkmaz et al. [14] in their work tried to estimate the transit and operational time linked to BiCMOS transistor. They observed that approximately 50 µA of current flows through a 0.8 µm BiCMOS transistor when it changes state, and the switching (transit) time is about 0.4 ns.

12.3 CHALLENGES TO ADOPT GREEN COMPUTING

The researchers in the past have focused on increasing computing efficiency and lowering the cost associated to IT equipment and infrastructure services. Now infrastructure is becoming the bottleneck in IT environments and the reason for this is the growing computing needs, energy cost, and global warming. This shift is a great challenge for IT industry. Therefore, now researchers are focusing on the cooling systems, power consumption, and data center space. At one extreme, it is the processing power that is important to business and on the other extreme it is the drive, challenge of creating an environment-friendly system, and infrastructure limitations.

12.4   IMPLEMENTATION OF GREEN COMPUTING

The consumption of energy sources has a negative impact on the environment. The computing technology raises several environmental problems like global warming, pollution, toxicity, occupational and health hazards. Computers are continuously using a large amount of power and consequently regular cooling energy is needed to counteract this power usage. There are three basic causes toward implementation of green technology approach. Initially, to save cost, it lowers the energy consumption and thereby the power bills by making IT systems more efficient. Also, to reduce environmental problems, it applies methodologies for manufacturing computers with the use of nontoxic materials. Finally, to minimize the energy consumption, it adopts effective approaches to use energy-efficient products. Citing an example of energy-efficient product, Wi-Fi scheduler, allows managing and scheduling as well as configuring the specific tasks, while the wireless network is switched toward saving energy as well as enhancing security measures. It allows switches to automatically detect link status and reduce power usage of ports that are idle. Switch can detect whether a device is connected to each port. If no device is connected to a port, the switch will automatically power down that port. Another example can be the green laptop which will be an innovative approach as well as eco-friendly. It should be recyclable, nontoxic, and disposable. It will consume less power with LED-backlit displays. It will also support cent percent recyclable packaging. It can be designed in many forms like recyclable paper laptop, bamboo laptops, etc. The technologies associated with these types of laptops are based on green technology from their designing and production to recycling and disposal. In addition to that it will also reduce carbon dioxide emission. The ratio of CPU utilization may be easily approximated within the stipulated time. In such case, mostly the modern operating systems provide the facility to report the CPU utilization, and also many embedded development systems provide the desired capability. So, as the two architectures have the same throughput where one consumes less energy than the other, choosing the former can reduce the operational cost.

12.5   CONCLUSION

In this chapter, the basic concepts of green computing and its benefits to sustainable environment are discussed. The use of a green computer may be linked to eradication of many toxic materials. Green computers are an appropriate application of green computing. The approach in this case can estimate the consumption of energy in the system while running on processor. It can be used to complement existing software architecture design and analysis methods. As the concept of green computing becomes more important, the scheduled technique will provide valuable information to software architects and designers.

REFERENCES

1. www.ijetae.com/files/Volume3Issue1/IJETAE_0113_56.pdf
2. Climate Savers Computing Initiative (2010) Retrieved http://www.climatesavers computing.org.

3. Soomro, T. R., Wahba, H., Perspectives of cloud computing: An overview, *14th International Business Information Management Association (IBIMA) Conference on Global Business Transformation through Innovation and Knowledge Management*, Istanbul, Turkey, 23–24 June 2010. http://www.ibima.org/TR2010/papers/soo.html.

4. Berl, A., Gelenbe, E., Di Girolamo, M., Giuliani, G., Energy-efficient cloud computing. *The Computer Journal*, Vol. 53, Issue 7, pp. 1045–1051, 2009. doi:10.1093/comjnl/bxp080, http://comjnl.oxfordjournals.org/content/53/7/1045.short?rss=1.

5. Userful, Userful is the Green Solution: reduce CO@ emissions and electronic waste, 2011. http://www2.userful.com/green-pcs.

6. Research reveals environmental impact of Google searches. Fox News. 2009-01-12. Retrieved 2009-01-15.

7. Developing Green Software, 2011, by Dr. Bob Steigerwald and Abhishek Agrawal Software and Services Group, Intel Corporation, Folsom, CA, USA.

8. Green Computing: Go Green and Save Energy, July 2013 by Mrs. Sharmila Shinde, Mrs. Simantini Nalawade, Mr. Ajay Nalawade in *International Journal of Advanced Research in Computer Science and Software Engineering*.

9. Guenach, M., Nuzman, C., Maes, J., Peeters, M., Trading off rate and power Consumption in DSL systems, In *Proceedings of IEEE GLOBECOM Workshop on Green Communication*, Lexington, Dec. 2009, pp. 1–5.

10. Curtis-Maury, M., Singh, K., McKee, S.A., Blagojevic, F., Nikolopoulos, D.S., de Supinski, B.R., Schulz, M., Identifying energy-efficient concurrency levels using machine learning, In *Proceedings of IEEE International Conference on Cluster Computing*, USA, Sept. 2007, pp. 488–495.

11. Schmidt, D., Stal, M., Rohnert, H., Buschmann, F., *Pattern-Oriented Software Architecture, Patterns for Concurrent and Networked Objects*. Wiley, Hoboken, NJ, 2000.

12. Hovland, F., Lilja, T., Martensson, A., New software platform for energy management, In *Proceedings of the 21st International Telecommunication Energy Conference*, Piscataway, NJ, June 1999, pp. 1–6.

13. Shirazi, J., *Java Performance Tuning*, O'Reilly. Sebastopol, CA, 2000.

14. Korkmaz, P., Akgul, B.E.S., Palem K.V., Characterizing the behavior of a probabilistic CMOS switch through analytical models and its verification through simulations. CREST Technical Report, No. TR-05-08-01, Georgia Institute of Technology, Aug 2005.

13 Voltage Sag Mitigation Using Transformerless Dynamic Voltage Restorer

Pradyumna Kumar Sahoo
GIET

Devidarshinee Pradhan and
Prasanta Kumar Satpathy
CET

Prateek Kumar Sahoo
SOA University

CONTENTS

13.1 INTRODUCTION

Among many types of disturbances that can appear in power system, voltage sags can lead to the highest level of undesirable impact on sensitive loads. Short circuit faults, starting up of large loads such as induction motors are the major causes of this voltage sag. For the compensation of the amount of voltage sag, dynamic voltage restorer (DVR) is designed. It is also proposed to attenuate the problem of series compensation. It is typically installed in a distribution system and its main function is to build up the amount of voltage needed to compensate the voltage sags or dips. It consists of energy storage device which are made up of capacitor banks, batteries, or flywheels.

The main objective of the DVR is to protect the sensitive loads from all supply-side voltage disturbances like voltage sag, voltage swell, voltage flicker, harmonics, etc. A voltage in required magnitude, frequency, and phase is injected by the DVR in order to nullify the disturbances. Either by using a single three-phase transformer or by three single-phase transformers, the compensating voltage is injected in series to the supply. Only reactive power injection is required to compensate the voltage sag to a certain extend. Active power injection is required when there are transients due to system disturbance. Reactive power can be generated by the DVR by its own, but in order to supply real power it needs energy storage devices like battery, capacitor, or flywheel. Transformerless DVR is better than the conventional one as it removes the voltage drop, harmonic loss, and phase shift. As the transformer operates at line frequency, it can be eliminated and hence the size and cost of the DVR can also be minimized.

13.2   STRUCTURE OF DVR

The conventional DVR circuit generally consists of the converter, energy storage device, filter, injection transformer, and a by-pass switch. The main objective of DVR is to compensate the voltage disturbances in the supply. It detects the amount of voltage disturbances in the supply and injects the required voltage to the supply in order to maintain a constant load voltage. Among the power quality problems in case of voltage sag, the reactive power is supplied to the load by the DVR and in case of voltage swell the reactive power is absorbed from the power supply. The voltage source converter (VSC), used in the DVR, is a power electronic device consisting a storage device which can generate a sinusoidal voltage at any required frequency, magnitude, and phase angle. The capacitor used in the VSC reduces the variations in output voltage. In the application of DVR, the VSC is used to replace the supply voltage or to generate the part of supply voltage which is missing. The main purpose of storage devices is to supply the necessary energy to the VSC via a DC-link for the generation of injected voltages. The different types of energy storage devices are like lead-acid batteries, superconducting magnetic energy storage, super capacitors, etc. The main task of the filtering unit is to keep the harmonic voltage content generated by the VSC to the permissible level. The higher-order switching harmonics generated by the pulse width Modulation (PWM) and VSC can improve the quality of the energy supply. Inverter side and line side filtering are the basic types of filtering schemes.

Figure 13.1 shows the schematic diagram of conventional DVR. The basic function of injection transformer is to boost the voltage generated by the vertical shift impactor (VSI). It serves the purpose of isolating the load from the system (VSC and control mechanism). It connects the DVR to the distribution network via the HV-windings and transforms and couples the injected compensating voltages generated by the VSCs to the incoming supply voltage.

13.3   TRANSFORMERLESS DVR

In case of the tansformerless DVR, the injection transformer is excluded. The schematic diagram of proposed DVR is shown in Figure 13.2. In the case of conventional three-phase DVR based on three single-phase VSIs, diagonally opposite switches in

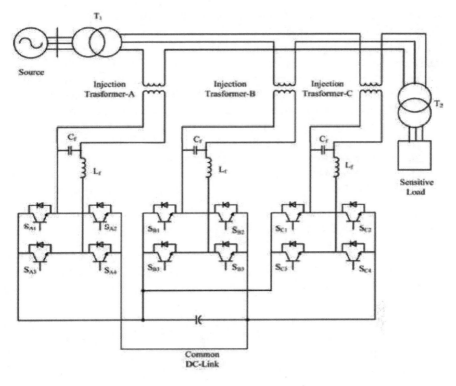

FIGURE 13.1 Schematic diagram for conventional DVR.

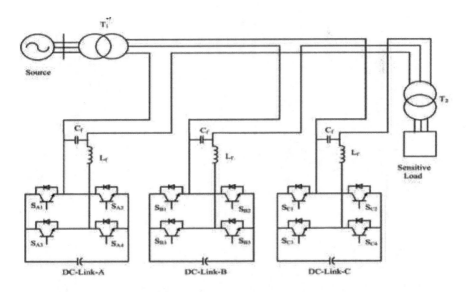

FIGURE 13.2 Schematic diagram of transformerless DVR with separate DC-links.

the VSIs will be on/off simultaneously. Conduction overlapping may occur within the three single-phase inverters. In case of conventional DVR, the risk of short circuit is less as an isolation transformer comes in between the phases. But when the transformer is removed in the case of transformerless DVR, the overlapping switches will make a short circuit between the phases and DC-link. This problem is solved by using separate DC-link for each of the single-phase inverters as shown in Figure 13.2. The maximum voltage injected by a transformerless DVR is decided entirely by the DC-link voltage. By using cascaded inverters, the shortcoming can be overcome. The major drawback of the conventional DVR is the use of a bulky and expensive injection transformer, which contributes to the total losses of the DVR. From the operation and maintenance point of view, the transformer also adds complexity to DVR system.

13.4   DVR CONTROL SCHEME

The control block using synchronous reference frame theory to control the DVR is shown in Figure 13.3. The load voltage V_L and source voltage V_S are used to produce the PWM pulses for the VSI.

$$\begin{bmatrix} V_{Lq} \\ V_{Ld} \\ V_{Lo} \end{bmatrix} = \frac{2}{3} \begin{bmatrix} \cos\theta & \cos\left(\theta - \frac{2\Pi}{3}\right) & \cos\left(\theta + \frac{2\Pi}{3}\right) \\ \sin\theta & \sin\left(\theta - \frac{2\Pi}{3}\right) & \sin\left(\theta + \frac{2\Pi}{3}\right) \\ \frac{1}{2} & \frac{1}{2} & \frac{1}{2} \end{bmatrix} \begin{bmatrix} V_{La} \\ V_{Lb} \\ V_{Lc} \end{bmatrix} \quad (13.1)$$

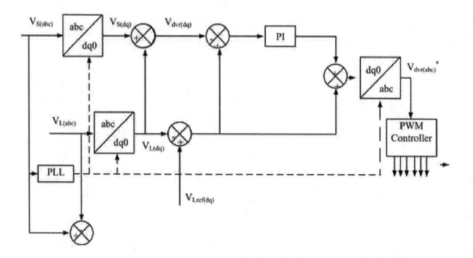

FIGURE 13.3   Synchronous reference frame method used for the control of the DVR.

TABLE 13.1

DVR Parameters

Description	Abbreviation	Value
Three phase line to line voltage	V_s	440 V
Active load power	P	7.9 kW
Reactive load power	Q	0.12 kVA

The reference load voltages $\left(V_{L_a}{}^*, V_{L_b}{}^*, V_{L_c}{}^*\right)$ and the source voltages are also converted to the rotating reference frame in the same manner. The DVR voltages can be expressed in the *dq0* frame as

$$V_{D_d} = V_{S_d} - V_{L_d} \tag{13.2}$$

$$V_{D_q} = V_{S_q} - V_{L_q} \tag{13.3}$$

And the reference DVR voltages in the *dq0* frame can be expressed as

$$V^*{}_{D_d} = V^*{}_{S_d} - V_{L_d} \tag{13.4}$$

$$V^*{}_{D_q} = V^*{}_{S_q} - V_{L_q} \tag{13.5}$$

Two proportional-integral controllers are used to regulate the error between actual and reference DVR voltage. Inverse Park's transformation is used to get reference DVR voltages in *abc* reference frame. Gating pulses for the VSI are generated using reference DVR voltage and actual DVR voltage (Table 13.1).

$$\begin{bmatrix} V^*{}_{dvr_a} \\ V^*{}_{dvr_b} \\ V^*{}_{dvr_c} \end{bmatrix} = \frac{2}{3} \begin{bmatrix} \cos\theta & \sin\theta & 1 \\ \cos\left(\theta - 2\Pi/3\right) & \sin\left(\theta - 2\Pi/3\right) & 1 \\ \cos\left(\theta + 2\Pi/3\right) & \sin\left(\theta + 2\Pi/3\right) & 1 \end{bmatrix} \begin{bmatrix} V^*{}_{D_q} \\ V^*{}_{D_d} \\ V^*{}_{D_0} \end{bmatrix} \tag{13.6}$$

13.5 SIMULATION RESULTS

A three-phase balanced source is used for the simulation study. The load is considered to be purely resistive. The voltage sag is generated creating a resistive fault at the source side. MATLAB®/Simulink is used for the simulation studies. Figure 13.4 shows the load reference voltage of the DVR. A voltage sag is observed in-between 4 and 6 s, as shown in Figure 13.5. And in Figure 13.6, the voltage injected by the DVR is shown. It can be observed from Figure 13.7 that the load voltage is maintained at reference voltage amplitude during the sag condition by the injection of required voltage.

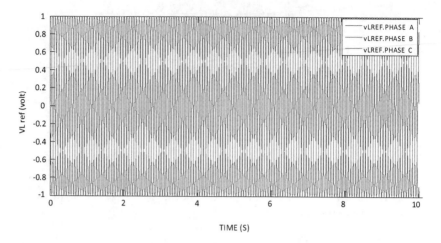

FIGURE 13.4   Load reference voltage.

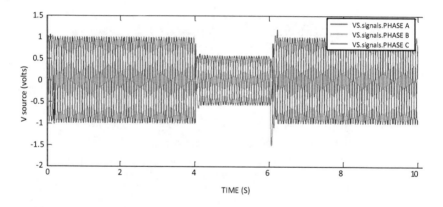

FIGURE 13.5   Source voltage with sag.

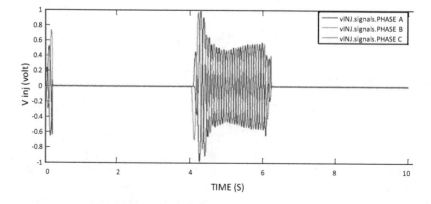

FIGURE 13.6   Injected voltage by the DVR.

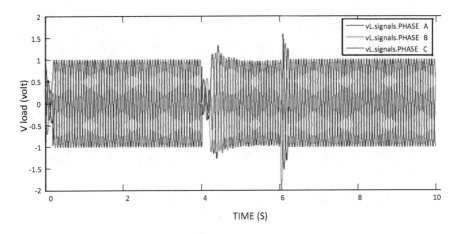

FIGURE 13.7   Load voltage with DVR.

13.6   CONCLUSION

In this chapter, we find that by the elimination of the injection transformer, large cost reduction is possible which makes it more attractive. Compensation techniques of power electronic devices like DVR are also presented along with the feasibility of introducing transformerless DVR. The design and applications of DVR for voltage sags and comprehensive results are presented. Unit vectors are used to find out the reference load voltage. Separate DC-links isolation for the DVR is proposed. The MATLAB results show that the proposed DVR can compensate the voltage sag satisfactorily. Hence, it may be concluded that the main advantage of the transformerless DVR over the conventional DVR is the reduction in cost and size.

REFERENCES

1. E.K.K. Sng, S.S. Choi, and D.M. Vilathgamua, Analysis of series compensation and DC-link voltage controls of a transformerless self-charging dynamic voltage restorer, *IEEE Trans. Power Delivery*, vol. 19, no. 3, pp. 1511–1518, Jul 2004.
2. E. Babaei, M.F. Kangarlu, and M. Sabahi, Mitigation of voltage disturbances using dynamic voltage restorer based on direct converters, *IEEE Trans. Power Delivery*, vol. 25, no. 4, pp. 2676–2683, Oct 2010.
3. N.H. Woodley, Field experience with dynamic voltage restorer (DVR MV) systems, *Proc. IEEE- PES Winter Meeting*, vol. 4, Singapore, pp. 2864–2871, Jan 2000.
4. T. Jimichi, H. Fujita, and H. Akagi, An approach to eliminating DC magnetic flux from the series transformer of a dynamic voltage restorer, *IEEE Trans. Ind. Appl.*, vol. 44, no. 3, pp. 806–816, May/Jun 2008.
5. H.K. AI-Hadidi, A.M. Gole, and D.A. Jacobson, A novel configuration for a cascade inverter- based dynamic voltage restorer with reduced energy storage requirements, *IEEE Trans. Ind. Appl.*, vol. 23, no. 2, pp. 881–888, Apr 2008.
6. S.S. Choi, B.H. Li, and D.M. Vilathgamua, Design and analysis of the inverter side filter used in the dynamic voltage restorer, *IEEE Trans. Power Delivery*, vol. 17, no. 3, pp. 857–864, Jul 2002.

7. D.M. Vilathgamua, H.M. Wijekoon, and S.S. Choi, Interline dynamic voltage restorer: a novel and economical approach for multi-line power quality compensation, *IEEE Trans. Ind. Appl.*, vol. 40, no. 6, pp.1678–1685, Nov/Dec 2004.

8. T. Jimichi, H. Fujita, and H. Akagi, A dynamic voltage restorer equipped with a high-frequency isolated DC-DC converter, *IEEE Trans. Ind. Appl.*, vol. 47, no. I, pp. 169–175, Jan/Feb 2011.

9. S.M. Silva, S.E. da Silveira, A.S. Reis, and B.J.C. Filho, Analysis of a dynamic voltage compensator with reduced switch-count and absence of energy storage system, *IEEE Trans. Ind. Appl.*, vol. 41, no. 5, pp.1255–1262, Sep/Oct 2005.

10. A. Prasai, and D.M. Deepak, Zero-energy sag correctors-optimizing dynamic voltage restorers for industrial applications, *IEEE Trans. Ind. Appl.*, vol. 44, no. 3, pp. 809–816, May/Jun 2008.

11. N. Mohan, T.M. Undeland, and W.P. Robbins, *Power Electronics: Converters, Applications and Design.* New York: Wiley, 1989.

14 Harmonics Reduction Using Active Power Filter with Hysteresis Current Control

Pati Sushil Kumar
GIET

Satpathy Suchismita
CVRP

CONTENTS

14.1 INTRODUCTION

Shunt active power filter (SAPF) is connected to the power system at the point of common coupling (PCC). Due to nonlinear load, the load current produces harmonics. The compensating current which is output of the active power filter (APF) is injected at PCC, so the harmonic cancellation takes place and the current between sources to PCC becomes sinusoidal in nature as shown in Figure 14.1.

The main circuit behind the APF is the control unit. The control unit is mainly divided into two parts: harmonic extraction technique and current modulator.

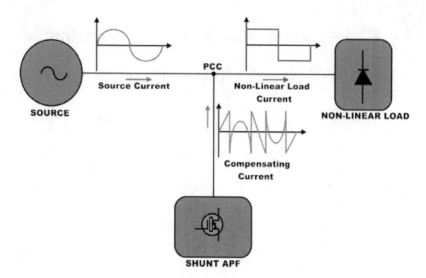

FIGURE 14.1

14.1.1 Harmonic Extraction

Harmonic extraction is the process in which reference current is generated by using the distorted waveform. Many theories are developed in these categories, for example, *p-q* theory (instantaneous active power theory), *d-q* theory, PLL controller, neural network, and sinusoidal harmonic controller. Here in this chapter, *d-q* theory is used due to its accuracy, robustness, and easy calculation.

14.1.2 Current Modulator (Gate Pulse)

Current modulator is mainly used to provide the gate pulse to the APF (VSI). There are many techniques used for giving the gate pulse. These are sinusoidal PWM, triangular PWM, hysteresis current controller (HCC), space vector modulation, space vector with HCC, etc. Here, HCC strategy is used.

14.1.3 Harmonics

"Harmonics" means a component with a frequency that is an integer multiple (the so-called order of harmonic *n*) of the fundamental frequency; the first harmonic is the fundamental frequency (50 or 60 Hz). The second harmonic is the component with frequency two times the fundamental (100 or 120 Hz), and so on, as shown in Figure 14.2. Harmonic distortion can be considered as a sort of pollution of the electric system which can cause problems if the sum of the harmonic currents exceeds certain limits. The utilization of electrical power mainly depends upon supply of power with controllable frequencies and voltages, whereas its generation and transmission take place at nominally constant levels. For these, some form of power conditioning or conversion is needed and normally implemented by power electronic

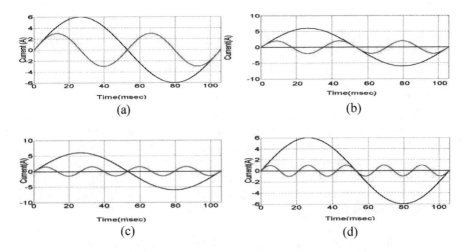

FIGURE 14.2 A sinusoidal waveform with fundamental frequency 50 Hz and its harmonics: (a) second (100 Hz), (b) third (150 Hz), (c) fourth (200 Hz), and (d) fifth (250 Hz).

circuitry that distorts the voltage and current waveforms. Therefore, the main source of harmonics in the power systems is the nonlinear loads.

14.1.4 TOTAL HARMONIC DISTORTION (THD)

The total harmonic distortion (THD) of a signal is a measurement of the harmonic distortion present and is defined as the ratio of the sum of the powers of all harmonic components to the power of the fundamental frequency. Harmonic distortion is caused by the introduction of waveforms at frequencies in multiplies of the fundamental, that is, the third harmonic is 3 multiplied by the fundamental frequency (150 Hz). THD is a measurement of the sum value of the waveform that is distorted.

$$\text{THD} = \frac{\sqrt{V_2^2 + V_3^2 + V_4^2 + \cdots + V_N^2}}{V_1} \tag{14.1}$$

The THD is a very useful quantity for many applications. It is the most commonly used harmonic index. However, it has the limitation that it is not a good indicator of voltage stress within a capacitor because that is related to the peak value of voltage waveform. The amount of harmonic distortion present in the system is generally notified using the terminology of "THD" and "TMD" which have been defined as per IEEE standards. The THD is defined as the root mean square value of the ratio of harmonic current to its fundamental load component of current.

14.2 SRF CONTROLLER

The synchronous reference frame (SRF) theory or d-q theory is based on time-domain reference signal estimation techniques. It performs the operation in steady state or transient state as well as for generic voltage and current waveforms. It allows controlling the APFs in real-time systems. Another important characteristic of this theory is the simplicity of the calculations, which involves only algebraic calculation. The basic structure of SRF methods consists of direct $(d\text{-}q)$ and inverse $(d\text{-}q)^{-1}$ Park transformations as shown in Figure 14.3, which allows the evaluation of a specific harmonic component of the input signals.

The reference frame transformation is formulated from a three-phase $a\text{-}b\text{-}c$ stationary system to the direct axis (d) and quadratic axis (q) rotating coordinate system. In $a\text{-}b\text{-}c$, stationary axes are separated from each other by $120°$ as shown in Figure 14.3. The instantaneous space vectors v_a and i_a are set on the a-axis, v_b and i_b are on the b-axis, similarly, v_c and i_c are on the c-axis. These three phase space vectors stationary coordinates are easily transformed into two axis $d\text{-}q$ rotating reference frame transformation. This algorithm facilitates deriving $i_d\text{-}i_q$ (rotating current coordinate) from three-phase stationary coordinate load current i_{La}, i_{Lb}, i_{Lc}.

$$\begin{pmatrix} i_{Ld} \\ i_{Lq} \\ i_{L0} \end{pmatrix} = \sqrt{\frac{2}{3}} \begin{pmatrix} \cos(\theta) & \cos(\theta - 2\pi/3) & \cos(\theta + 2\pi/3) \\ \sin(\theta) & \sin(\theta - 2\pi/3) & \sin(\theta + 2\pi/3) \\ 1/\sqrt{2} & 1/\sqrt{2} & 1/\sqrt{2} \end{pmatrix} \begin{pmatrix} i_{La} \\ i_{Lb} \\ i_{Lc} \end{pmatrix} \qquad (14.2)$$

The $d\text{-}q$ transformation output signals depend on the load current (fundamental and harmonic components) and the performance of the phase locked loop (PLL). The PLL circuit provides the rotation speed (rad/s) of the rotating reference frame, where ωt is set as fundamental frequency component. The PLL circuit provides the vectorized 50 Hz frequency and $30°$ phase angle followed by $\sin\theta$ and $\cos\theta$ for synchronization. The $i_d\text{-}i_q$ currents are sent through low pass filter for filtering the harmonic components of the load current, which allows only the fundamental frequency components. The low pass filter is a second-order Butterworth filter, whose cut-off frequency is selected to be 50 Hz for eliminating the higher-order harmonics. The proportional-integral (PI) controller is used to eliminate the steady-state error of the DC component of the d-axis reference signals. Furthermore, it maintains the

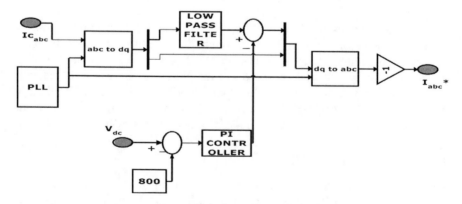

FIGURE 14.3   Synchronous $d\text{-}q\text{-}0$ reference frame–based compensation algorithm.

capacitor voltage nearly constant. The DC-side capacitor voltage of PWM-voltage source inverter is sensed and compared with desired reference voltage for calculating the error voltage. This error voltage is passed through a PI controller whose propagation gain (K_P) and integral gain (K_I) are 0.1 and 1, respectively (Figure 14.4).

14.3 HYSTERESIS CURRENT CONTROLLER

The hysteresis band current controller for SAPF can be carried out to produce the switching pattern of the inverter. There are various current control methods proposed for such APF configurations, but for quick current controllability and easy implementation hysteresis current control method has the highest rate among other current control methods.

Hysteresis band current controller has properties like robustness, excellent dynamics, and fastest control with minimum hardware (Figure 14.5).

The two-level PWM-voltage source inverter systems of the HCC are utilized independently for each phase. Each current controller directly generates the switching signal of the three phases. In the case of positive input current, if the error current $e(t)$ between the desired reference current $i_{ref}(t)$ and the actual source current $i_{actual}(t)$ exceeds the upper hysteresis band limit ($+h$), the upper switch of the inverter arm becomes OFF and the lower switch becomes ON, as shown in Figure 14.6.

FIGURE 14.4 *a-b-c* to *d-q*-0 transformation.

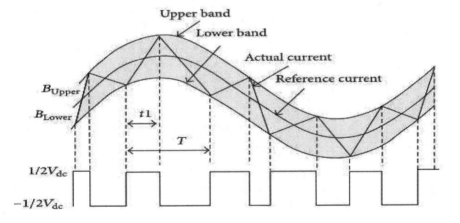

FIGURE 14.5 Hysteresis current band.

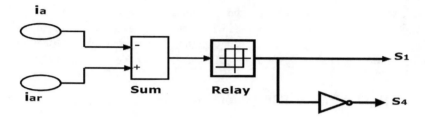

FIGURE 14.6   Hysteresis band current controllers.

14.3.1   ADVANTAGES OF HYSTERESIS PWM

- Excellent dynamic response
- Low cost and easy implementation

14.4   SIMULATION AND RESULTS

See Figure 14.7.

14.4.1   SIMULATION

Figure 14.8 shows the simulation model for a nonlinear load with three-phase supply and here APF is used. In Table 14.1, the system parameters considered for the study of APF for SRF controller with HCC are shown. The control part is used for generation of reference current (SRF controller) and providing gate signal (hysteresis current controller) to the PWM voltage source inverter. The nonlinear current is sensed by the current sensor and voltage across the capacitor is sensed by the voltage sensor, which is fed to the SRF controller. The reference current generated by SRF controller and the source current is fed to the current controller. Finally, the current controller is provided with the gate pulses to the PWM voltage source inverter. The output of

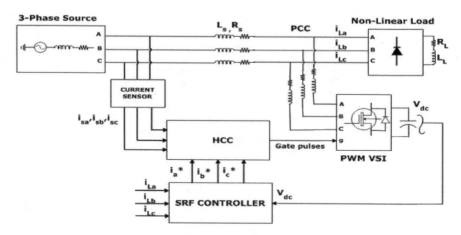

FIGURE 14.7   SRF with HCC-based APF implemented with PWM VSI configuration.

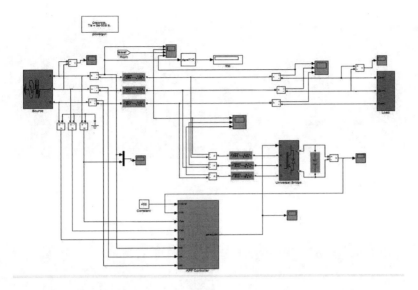

FIGURE 14.8   Simulation using APF with HCC.

TABLE 14.1
APF System Parameter

System Parameters	Value
Line voltage	440 V
Supply frequency	50 Hz
Source impedance: (resistance, Rs; inductance, Ls)	1 Ω, 0.1 μH
Nonlinear load under steady state: (resistance, Rs; inductance, Ls)	10 Ω, 100 μH
Filter impedance: (resistance, Rs; inductance, Ls)	1 Ω, 2.5 μH
DC side capacitance	1,400 μF
Reference DC voltage	800 V
Power converter	6 MOSFET/DIODE

the inverter i is called compensating current, which is injected to the power system at the PCC with opposite phase to the load current. The PI controller used inside the SRF controller for maintaining the capacitor voltage constant having propagation gain (K_P) and integral gain (K_I) are 0.1 and 1, respectively, and other system parameters are given in Table 14.1 (Figure 14.9).

14.4.2   RESULTS

In steady-state condition, the simulation time is taken as $t = 0$ to $t = 0.2$ s with constant load. The source voltage is shown in Figure 14.10. Load current is shown in Figure 14.11, which is highly nonlinear in nature with harmonic components.

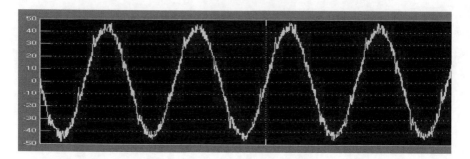

FIGURE 14.9 Source current.

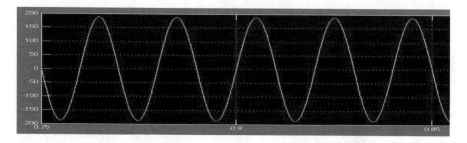

FIGURE 14.10 Source voltage.

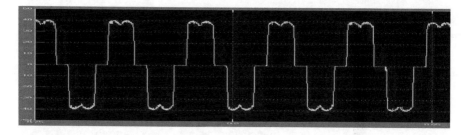

FIGURE 14.11 Load current.

The actual reference current for phase A is shown in Figure 14.12. This waveform is obtained from SRF controller. The APF supplies the compensating current to PCC, which is shown in Figure 14.13. The current after compensation is shown in Figure 14.14. It is clear from the figure that the waveform is sinusoidal with some frequency ripples.

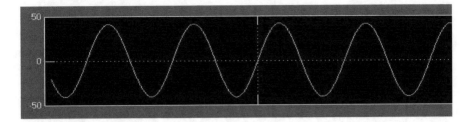

FIGURE 14.12   Reference current.

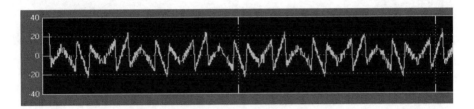

FIGURE 14.13   Compensating current.

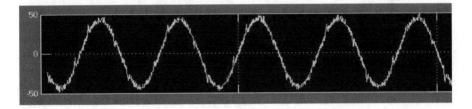

FIGURE 14.14   Source current after compensation.

14.4.3   THD CALCULATION

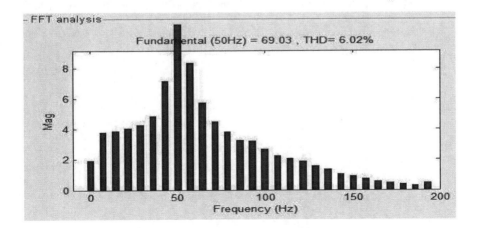

14.5   CONCLUSION

Here, in this chapter, it is shown with simulation that nonlinear load creates har-monics which is dangerous for the power system. APF with HCC is an option for better result. For switching signals to drive the VSI of the APF, the control strat-egy, namely, HCC is used. From the simulation result, it is observed that due to the compensating current supplied by the SAPF, the load current becomes sinusoidal. The THD is calculated to be 6.02%. So, the above control strategy is suitable for harmonic mitigation using APF.

REFERENCES

1. H.L. Jou, J.C. Wu, Y.J. Chang, and Y.T. Feng, A novel active power filter for harmonic suppression. *IEEE TPD*, Vol. 20, no. 2, 1507–1513, April 2005.
2. B. Singh, and K. Al-Haddad, A review of active filters for power quality improvement. *IEEE TIE*, Rourkela, Vol. 46, no. 5, 960–971, October 1999.
3. M. Kale, and E. Ozdemir, A novel adaptive hysteresis band current controller for shunt active power filter. *IEEE Conference, Control Applications, CCA*, Turkey, 2003.
4. H. Akagi, New needs in active filter for improving power quality, *International Conference on Power Electronics, Drives and Energy System/or Industrial Growth*, New Delhi, pp. 417–425, 1996.
5. C.T. Pan, An improved hysteresis current controller for reducing switching frequency. *IEEE TPE*, Vol. 9, no. I, 97–104, January 1994.
6. H.L. Jou, J.C. Wu, Y.J. Chang, and Y.T. Feng, A novel active power filter for harmonic suppression. *IEEE TPD*, Vol. 20, no. 2, 1507–1513, April 2005.
7. C.N. Bhende, S. Mishra, and S.K. Jain, TS-fuzzy-controlled active power filter for load compensation. *IEEE TPD*, Vol. 21, no. 3, 1456–1465, July 2006.
8. T.L. Lee, J.C. Li, and P.T. Cheng, Discrete frequency tuning active filter for power system harmonics. *IEEE TPE*, Vol. 24, no. 5, 202–208, May 2009.
9. S.R. Prusty, S.K. Ram, B.D. Subudhi, and K.K. Mahapatra, Performance analy-sis of adaptive band hysteresis current controller for shunt active power filter. *IEEE Conference, Emerging Trends in Electrical and Computer Technology*, St. Xavier's Catholic College of Engineering, Nagercoil, Kanayakumari, March 23–24, pp. 429–435, 2011.
10. S. K. Ram, S. R. Prusty, B.D Subudhi, and K.K Mahapatra, FPGA implementation of digital controller for active power line conditioner using SRF theory. *International Conference on Environment and Electrical Engineering (EEEIC-2011)*, Rome, Italy, May 8–11, 2011, IEEE.

Index

149